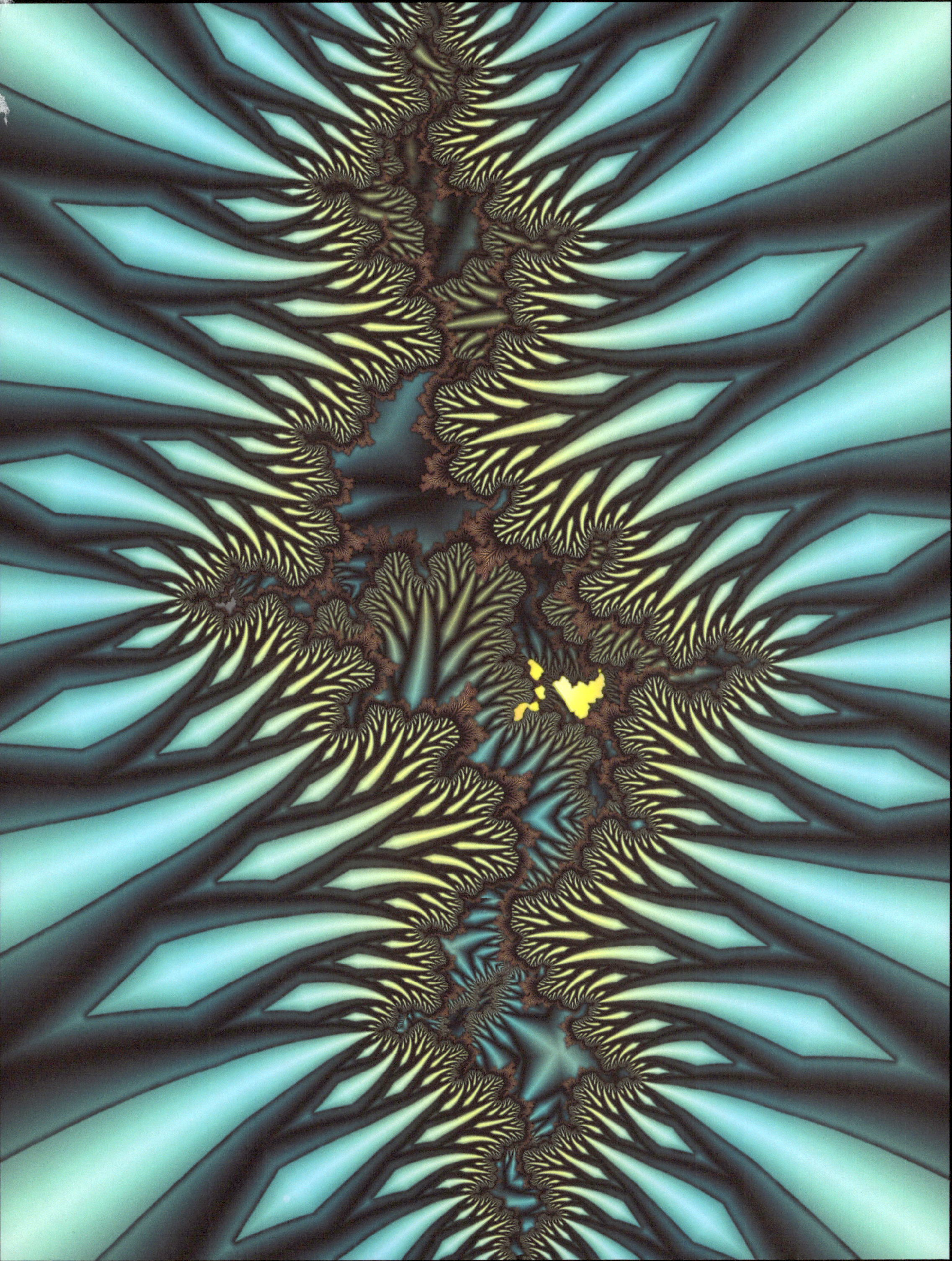

FRACTALS

Published in 2020 by Anamcara Press LLC
Author © 2020 by Jack Cleveland
Cover image & illustrations by Jack Cleveland
Cover & Book design by Maureen Carroll
Typeset in Trebuchet MS, Timeburner, Aviner, Ten Oldstyle.

Printed in the United States of America. This is a book for art lovers, designers, and art-loving techies everywhere. A coffee-table art book filled with lush art plates that speak to the senses, the fractal images within reflect the beauty and mystery of the natural world, and demonstrate the power of computer-aided design in creating original works of art.

ANAMCARA PRESS LLC

P.O. Box 442072, Lawrence, KS 66044

https://anamcara-press.com

Ordering Information:
Quantity sales. Special discounts are available on quantity purchases by corporations, associations, and others. For details, contact the publisher at the address above.
Orders by U.S. trade bookstores and wholesalers. Please contact Ingram Distribution.

Arts & Photography > Other Media > Digital
Science & Math > Mathematics > Geometry & Topology
Science & Math > Mathematics > Pure Mathematics > Fractals

Limited Collector's Edition *Fractals: Fractal Art* | 8 x 10.88 | 144 pages
A limited special edition art book, case-bound with 72 high-resolution quality plates beautifully produced on 80# satin finish paper, and individually shrink-wrapped to assure it is received in pristine condition, is available, along with select art prints, through the Anamcara Press website.

ISBN-13: 978-1-941237-29-8 (Limited Edition hardcover)
ISBN-13: 978-1-941237-58-8 (hardcover)
ISBN-13: 978-1-941237-60-1 (paperback)
ISBN-13: 978-1-941237-59-5 (eBook)

Library of Congress Control Number: 2020936650

FRACTALS

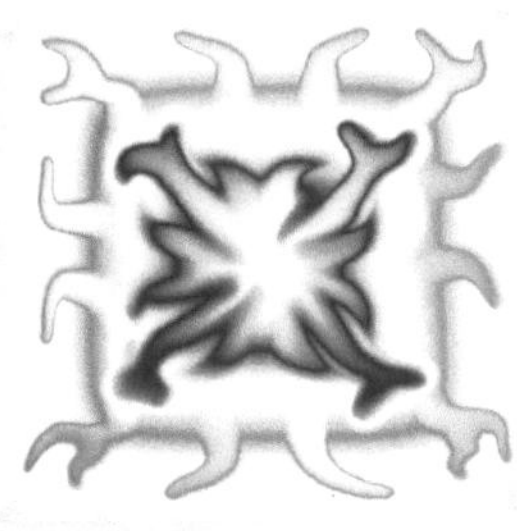

JACK CLEVELAND

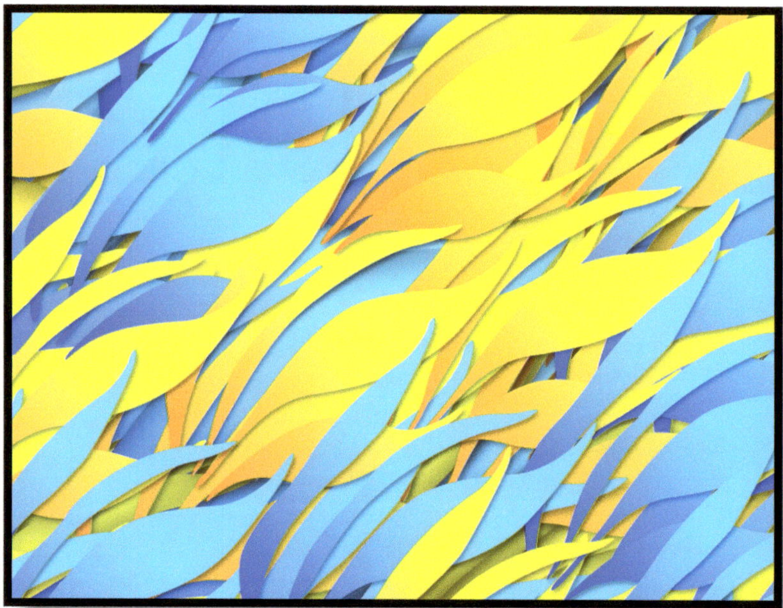

ON THE COVER

BIRDSOFAFEATHER, fractal image by Jack Cleveland

DEDICATION

To Robert

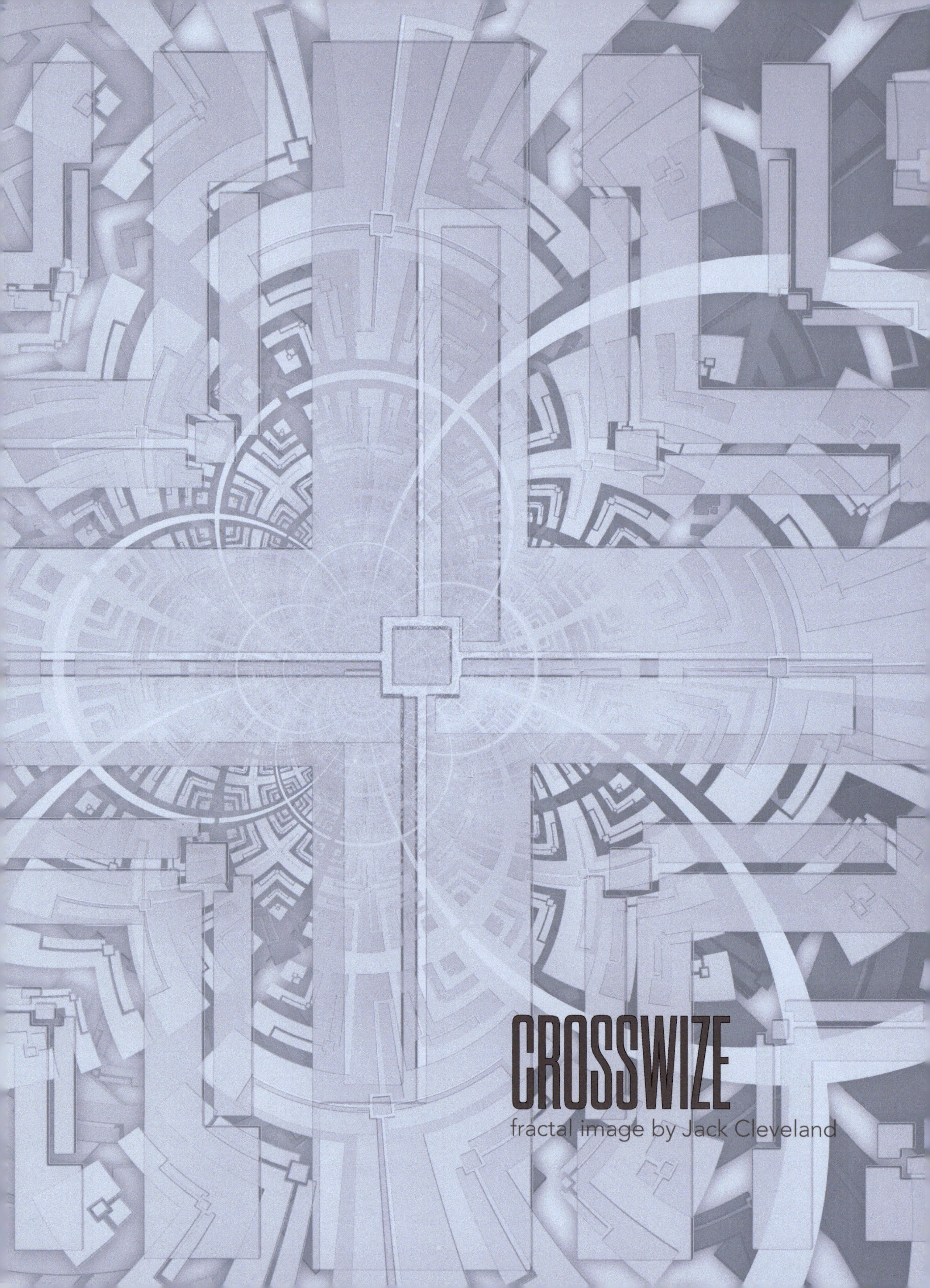
CROSSWIZE
fractal image by Jack Cleveland

Contents

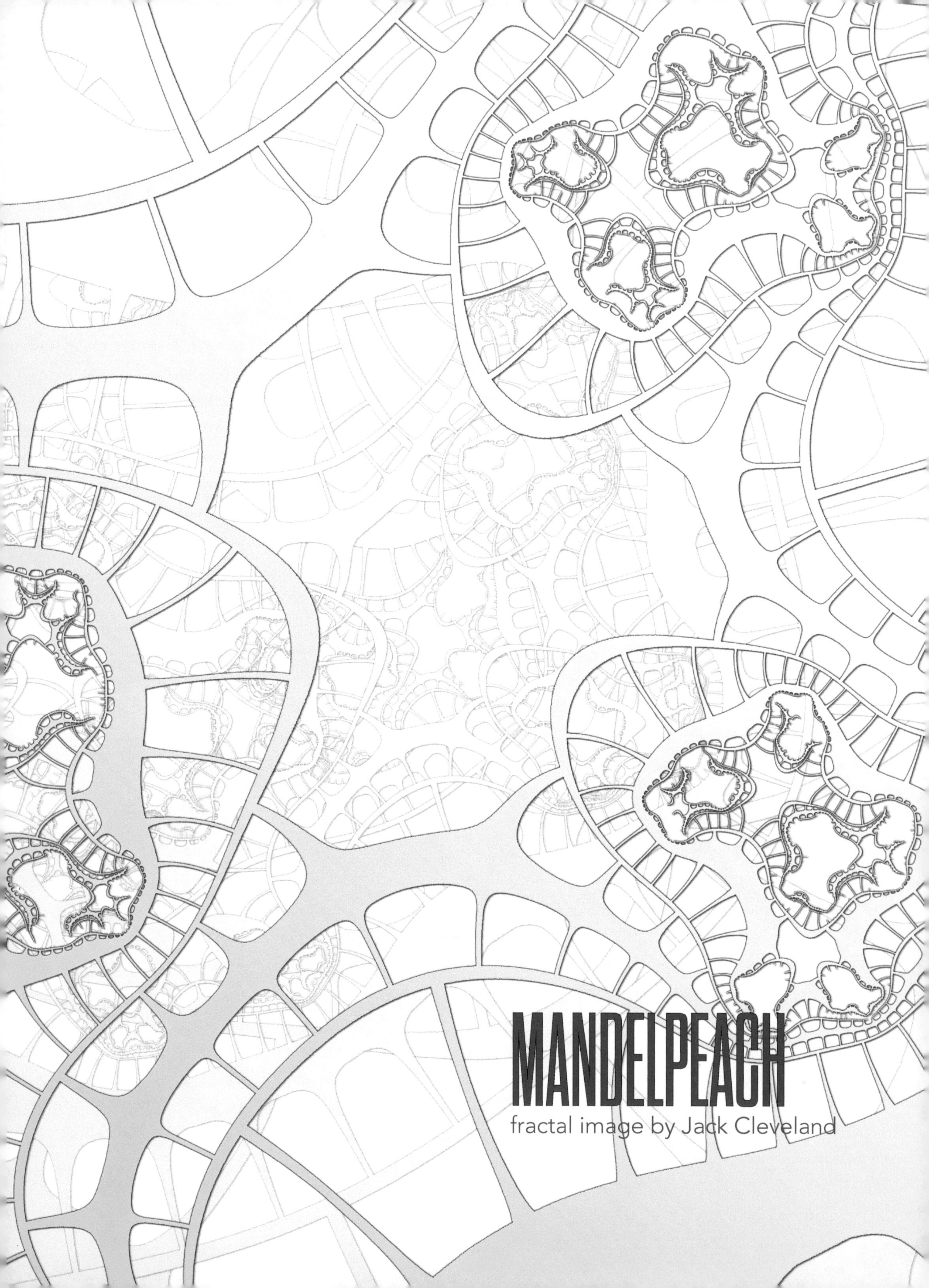
MANDELPEACH
fractal image by Jack Cleveland

FOREWORD

What is a fractal? A fractal is essentially a never-ending, repeating pattern. Fractal geometry can reflect and illuminate structures and patterns found in nature—from mountains, coastlines, and hurricanes to movements of the stock market over time. Fractal images are not made by the computer—they are not "computer-generated"—but rather are generated using computer graphics to create new and compelling works of art to represent dynamic phenomena not fully captured by classical mathematics.

This book is a compilation of fractal images created by Jack Cleveland. Publishing this book has been a labor of love, as Jack passed away before the book was completed, taking with him the secrets of how he created these marvelous images. We have endeavored to faithfully represent his vision, and encourage the reader to attempt to decipher the codes embedded in the images to create new worlds and insights. Perhaps, fractals can be seen to represent a metaphor for our chaotic times, as we attempt to understand the processes and structures which are manifest in the chaos and ineffable beauty of our ever-changing and dynamic natural world.

—Pat Cleveland

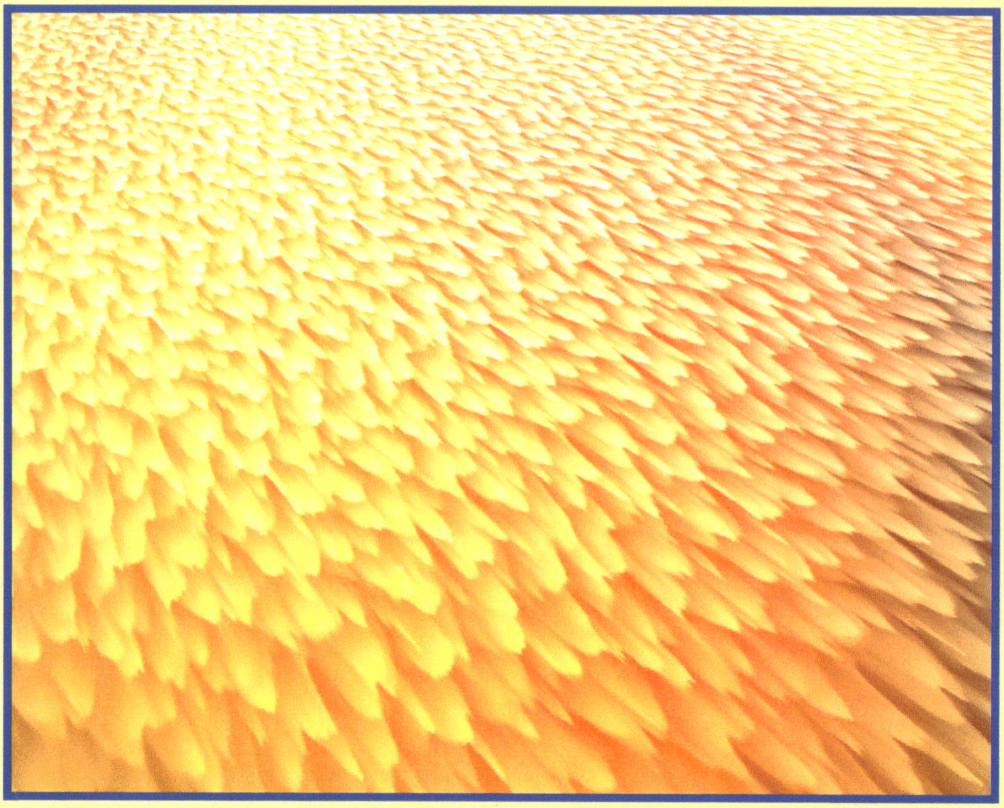

BIRDSOFAFEATHER I
fractal image by Jack Cleveland

INTRODUCTION

Benoit Mandelbrot (1924-2010) is considered the father of fractal geometry, indeed the inventor of the very word "fractal". His work made possible the visualization of geometry and transformed analysis of natural—and even financial—phenomena characterized by extreme variability and roughness. Not only did Mandelbrot's work have a radical impact on our understanding of information theory, economics and finance, nonlinear dynamics and geophysics, it dramatically revealed the underlying beauty inherent in mathematical laws governing turbulence and chaos, leading to the development of his concept of multifractals. The computer "pictures" he developed were alternative ways of representing new mathematical structures that came to be known as the Mandelbrot Set. This new branch of mathematics, in turn, led to the type of visual experimentation implemented by computers which is now ubiquitous in a wide range of applications.

> "In Fractals: Fractal Art a new category in digital art emerges. Jack Cleveland introduces us to his computer graphics-generated art forms that are geometric representations of specialized mathematical formulas, of which a great popular example is a Fibonacci Sequence. Before the advent of computers, scientists had been studying fractals for hundreds of years. As Benoit Mandelbrot's discoveries popularized fractal geometry, explorers and artists like Cleveland created shapes that are otherworldly yet visually arresting. His collection of art in Fractals is the product of diverse calculation formulas and coloring algorithms where he takes control. Cleveland's work can be classified as a new form of abstract art since they represent objective realities through layers of elements that achieve striking effects. As a fractal artist, Cleveland demonstrates how it can be unique, for it is the only medium that explores fractal structures through digital and classical aesthetics."
>
> —*Vincent Dublado, Readers' Favorite.*

JACK CLEVELAND
(1973–2020)

Cleveland's artwork clearly illustrates the power of Mandelbrot's revolutionary ideas for representing nature's complexity in all its beauty in a new mathematics embedded in computer-aided design. A student of environmental science and policy, he had a keen interest in maps, geographical information systems, geology and earth sciences, and his artwork reflects the overlapping realms of natural science and computer application and the underlying geometry of fractals in these unique pictures of which we imagine Mandelbrot would be proud.

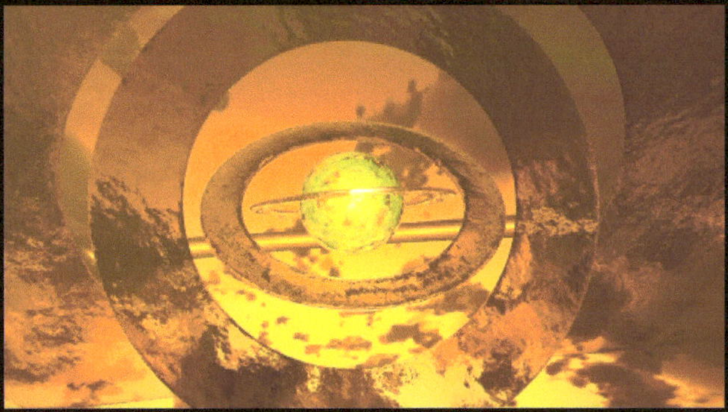

"He saw that infinite complexity could be described by simple rules...
He gave us order out of chaos, he gave us hope where there was none
His geometry succeeds where others fail.
So if you ever lose your way, a butterfly will flap its wings
From a million miles away, a little miracle will come to take you home."

"Mandelbrot Set" by Jonathan Coulton
https://www.jonathancoulton.com/

BLUEFIRE

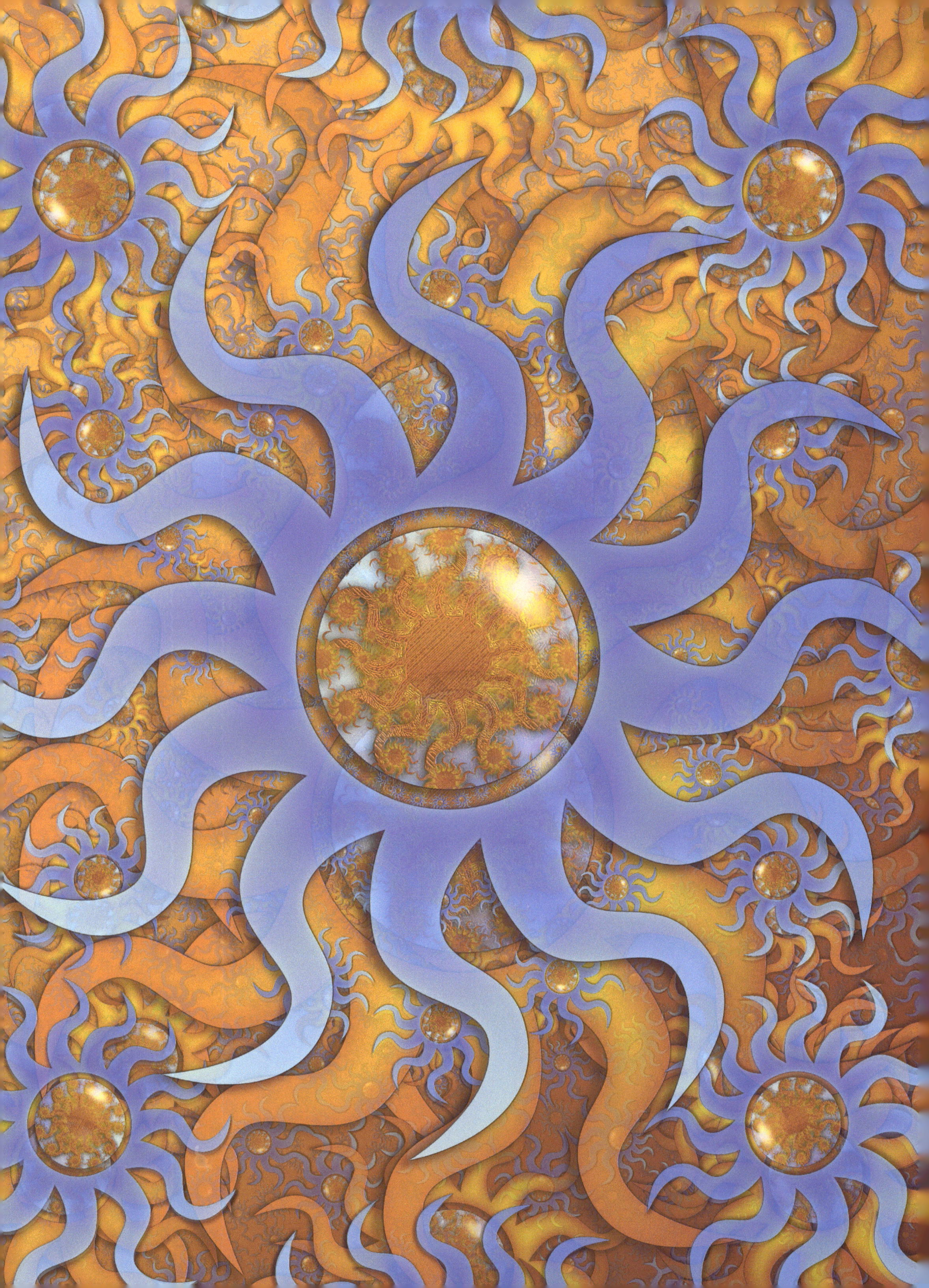

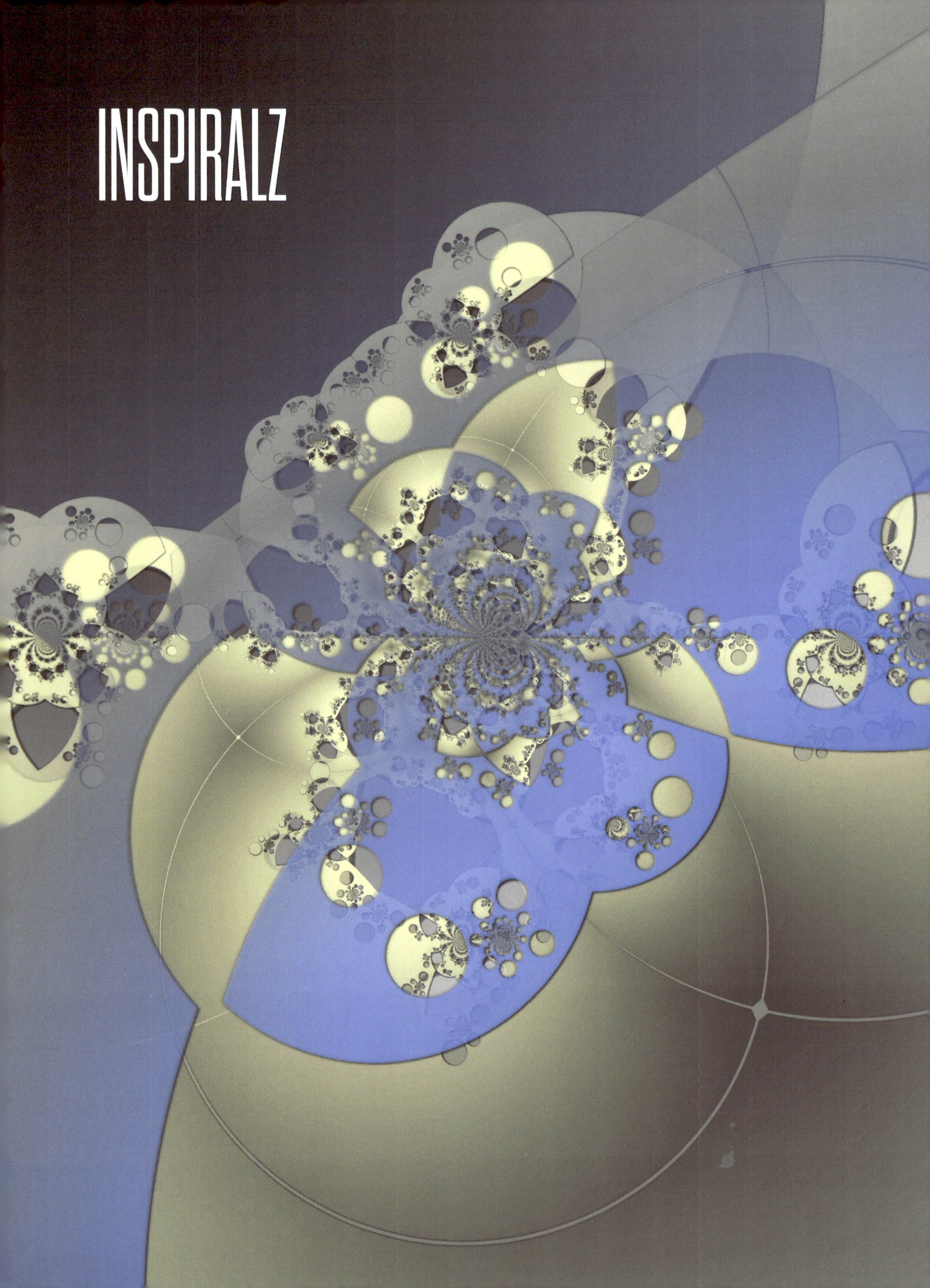
INSPIRALZ

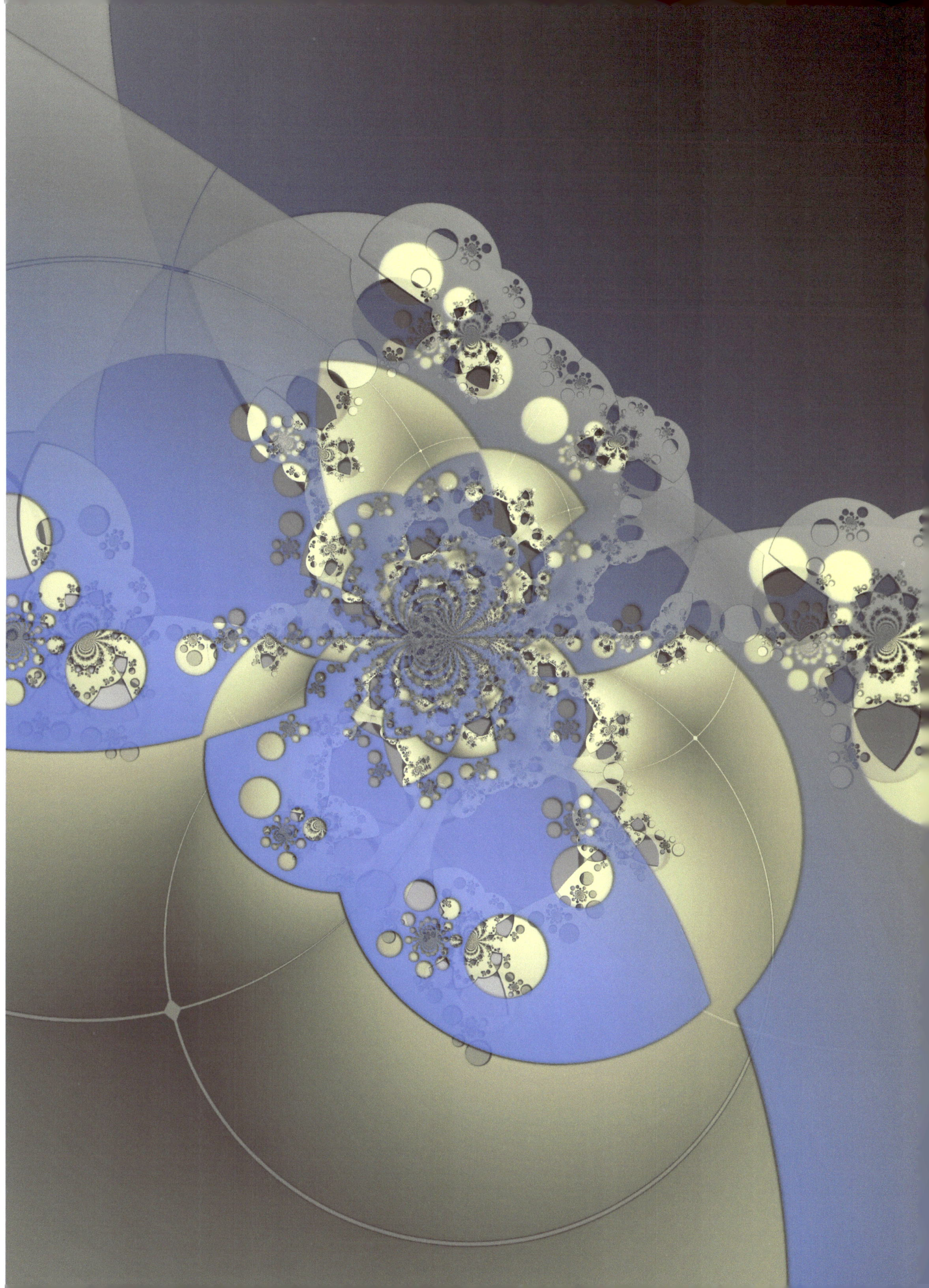

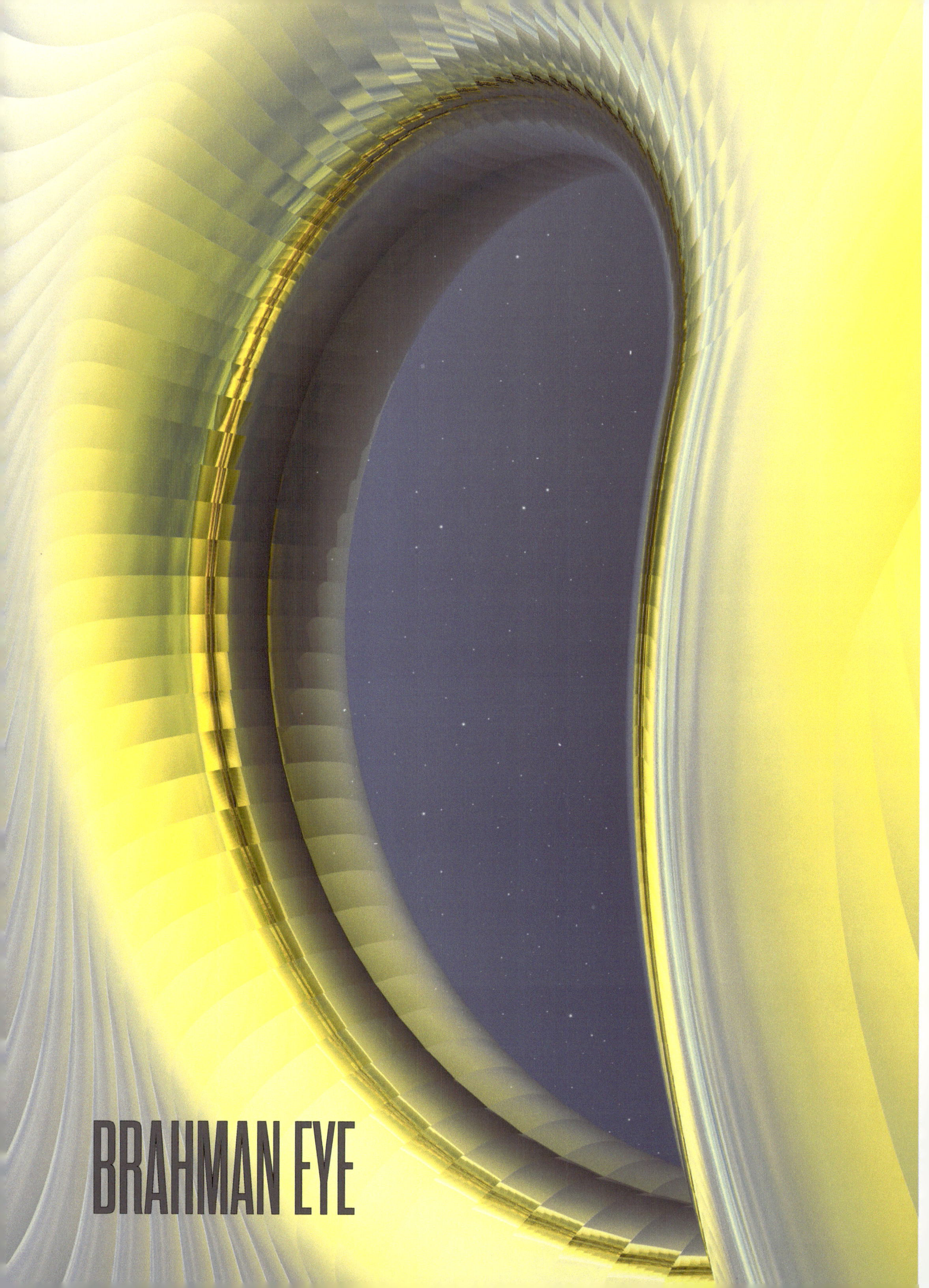
BRAHMAN EYE

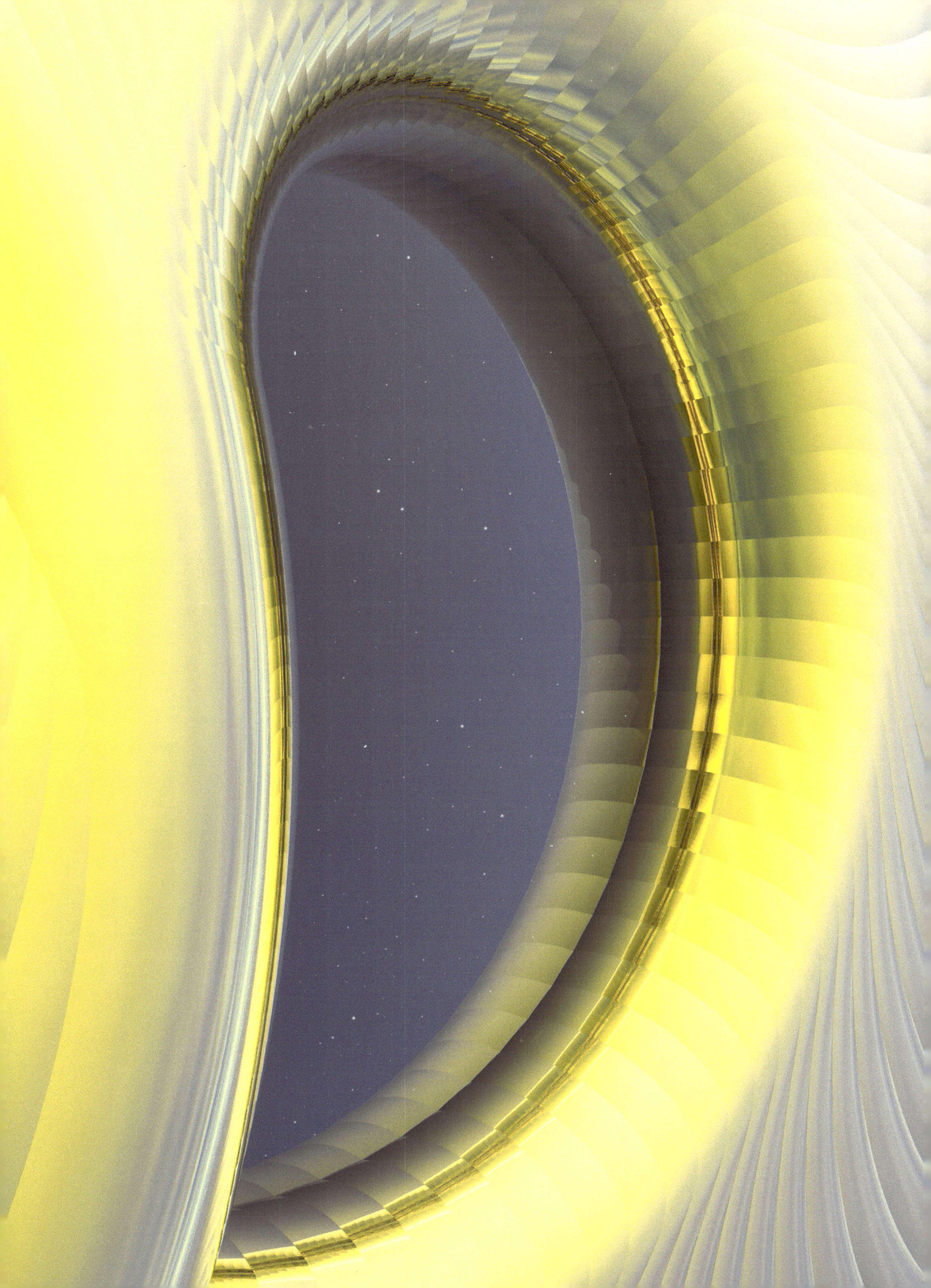

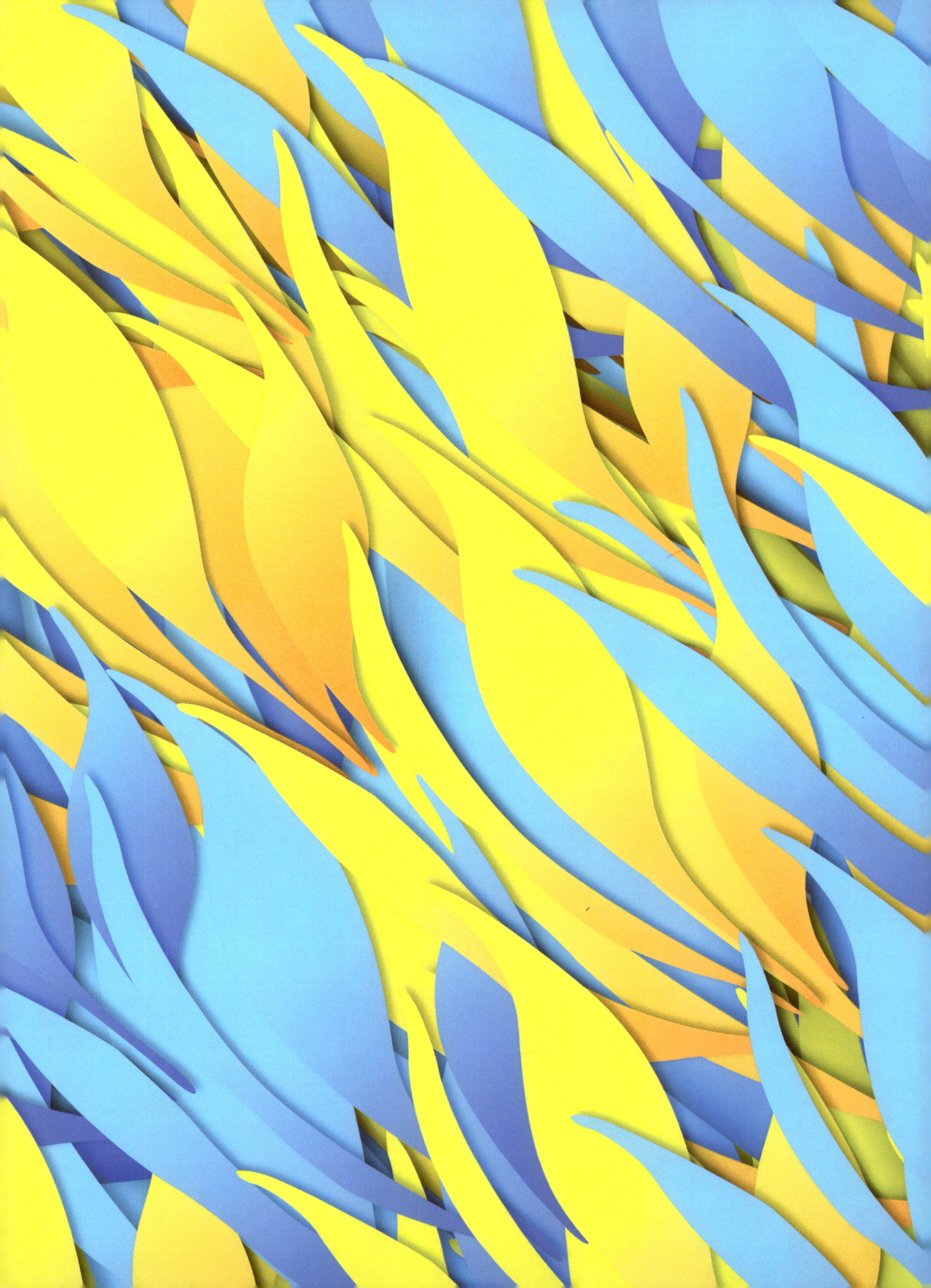

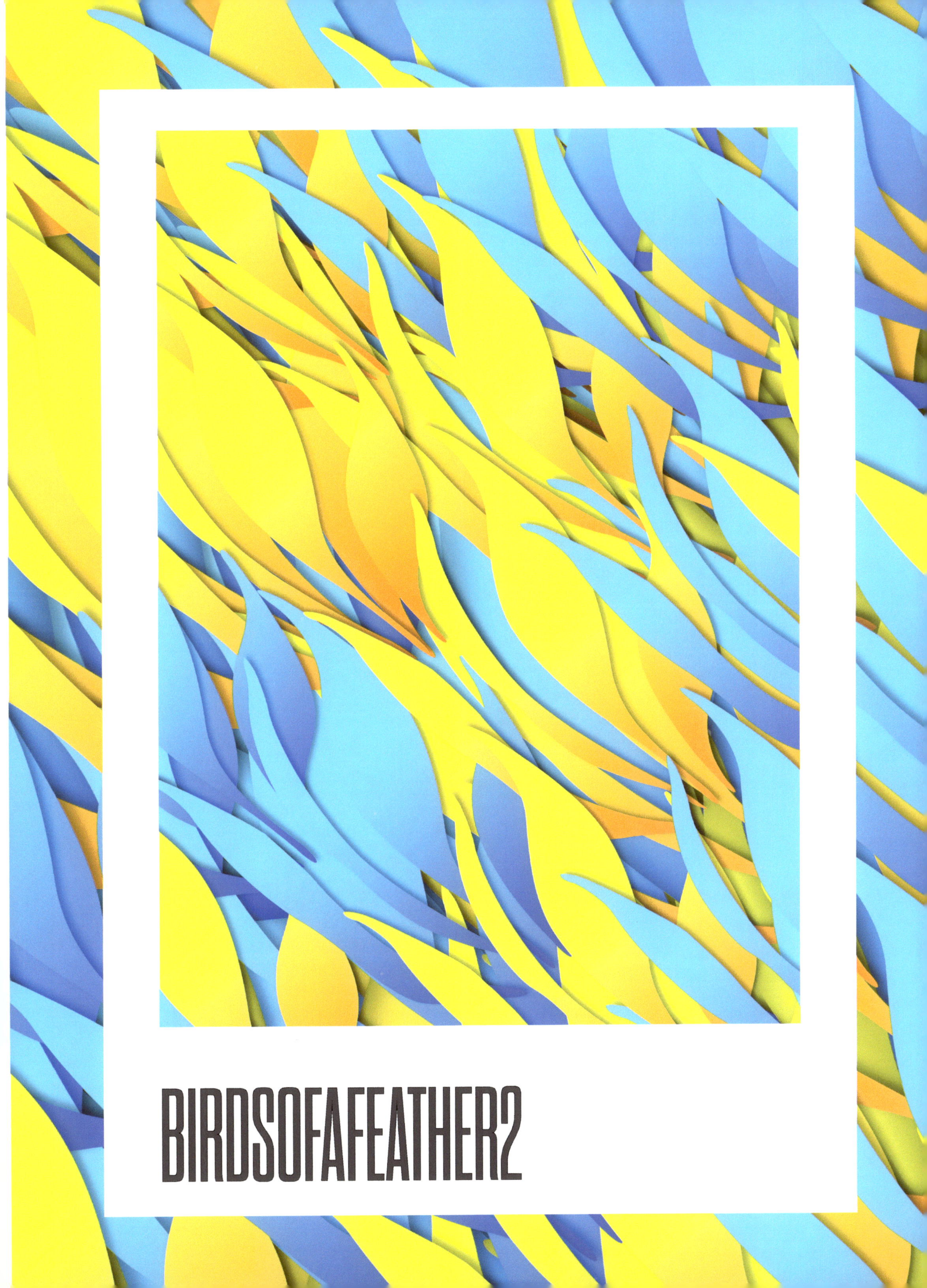
BIRDSOFAFEATHER2

KEYS

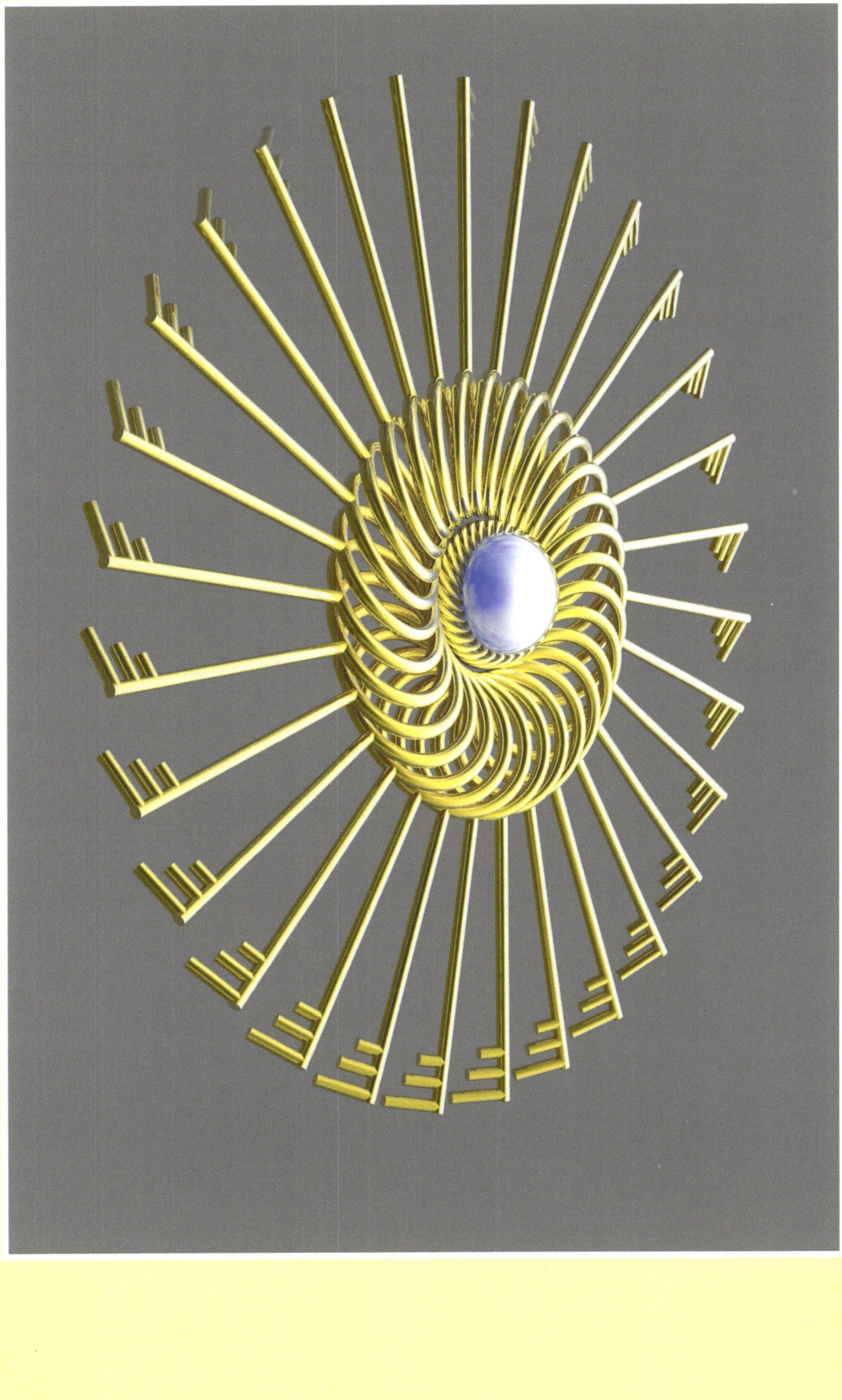

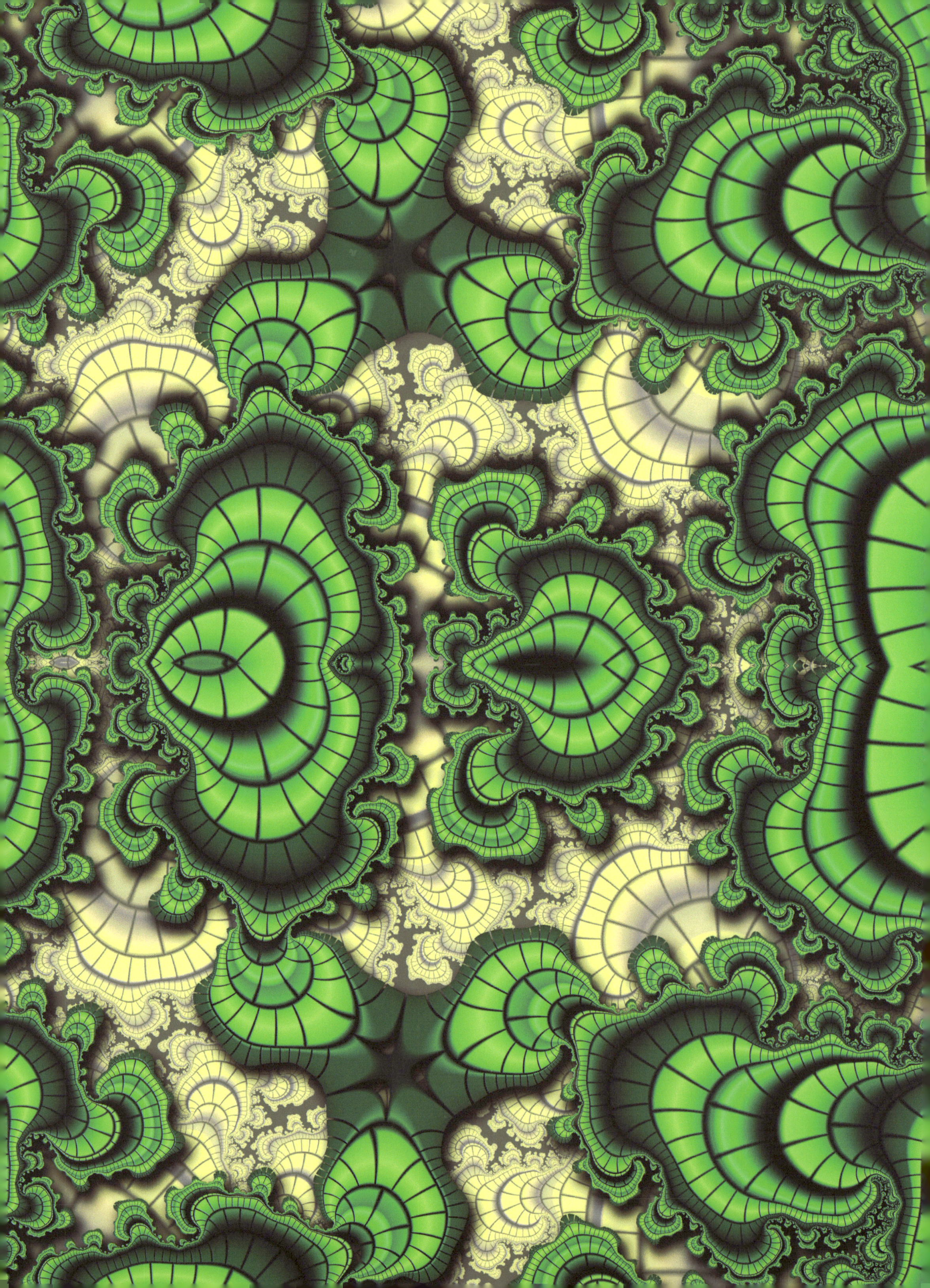

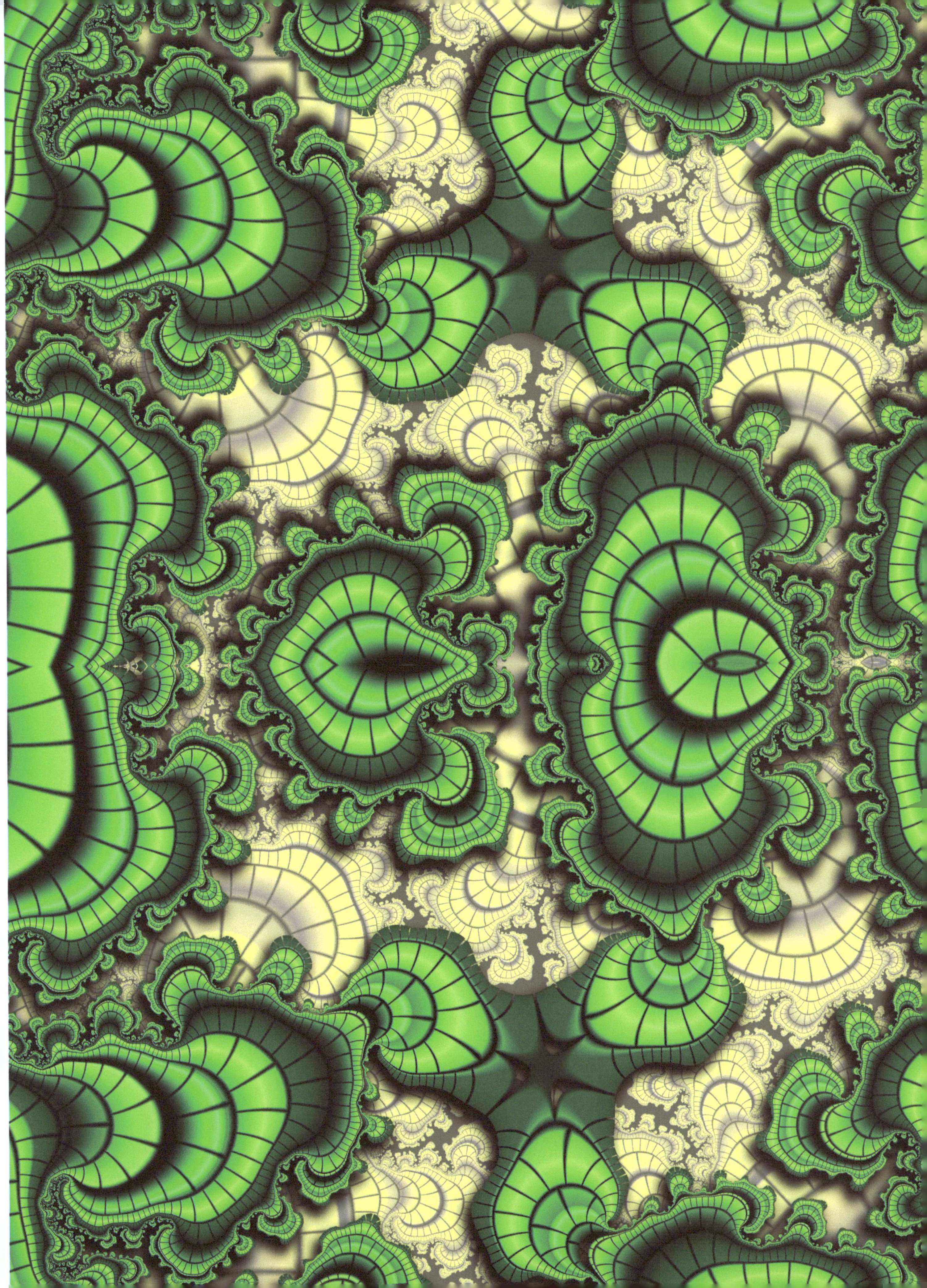

Previous pages:

LIFESPRINGS

PROTOS

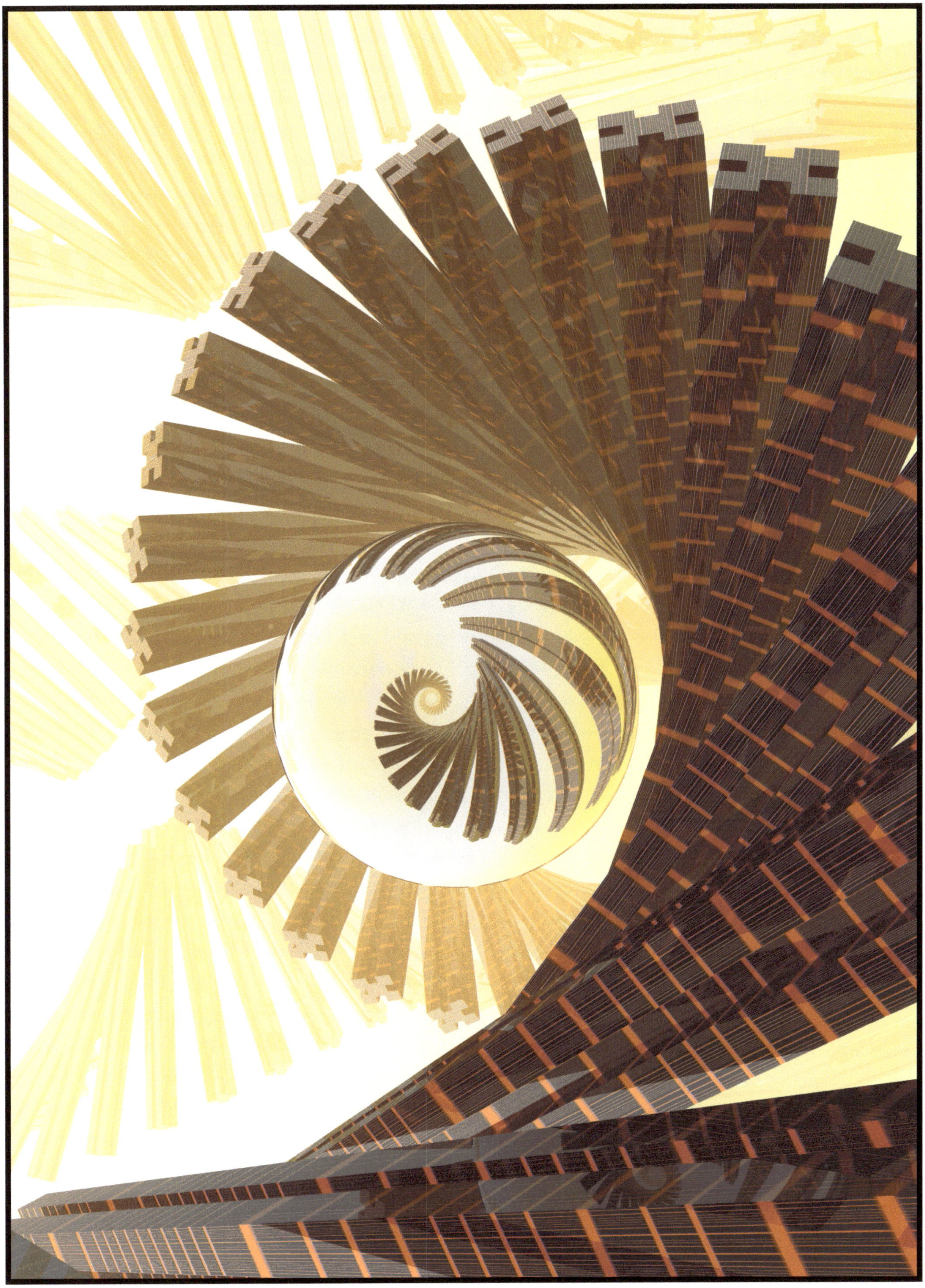

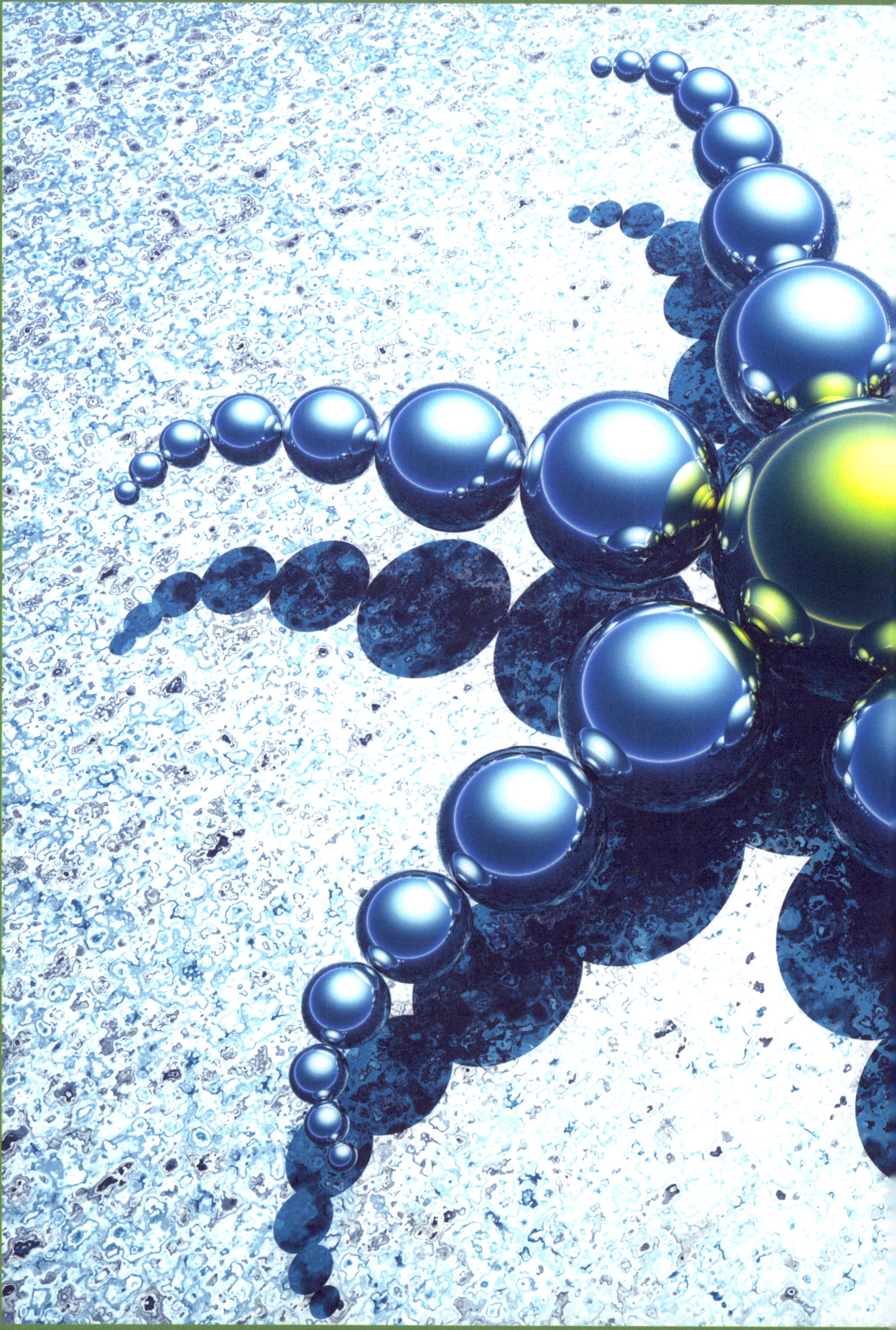

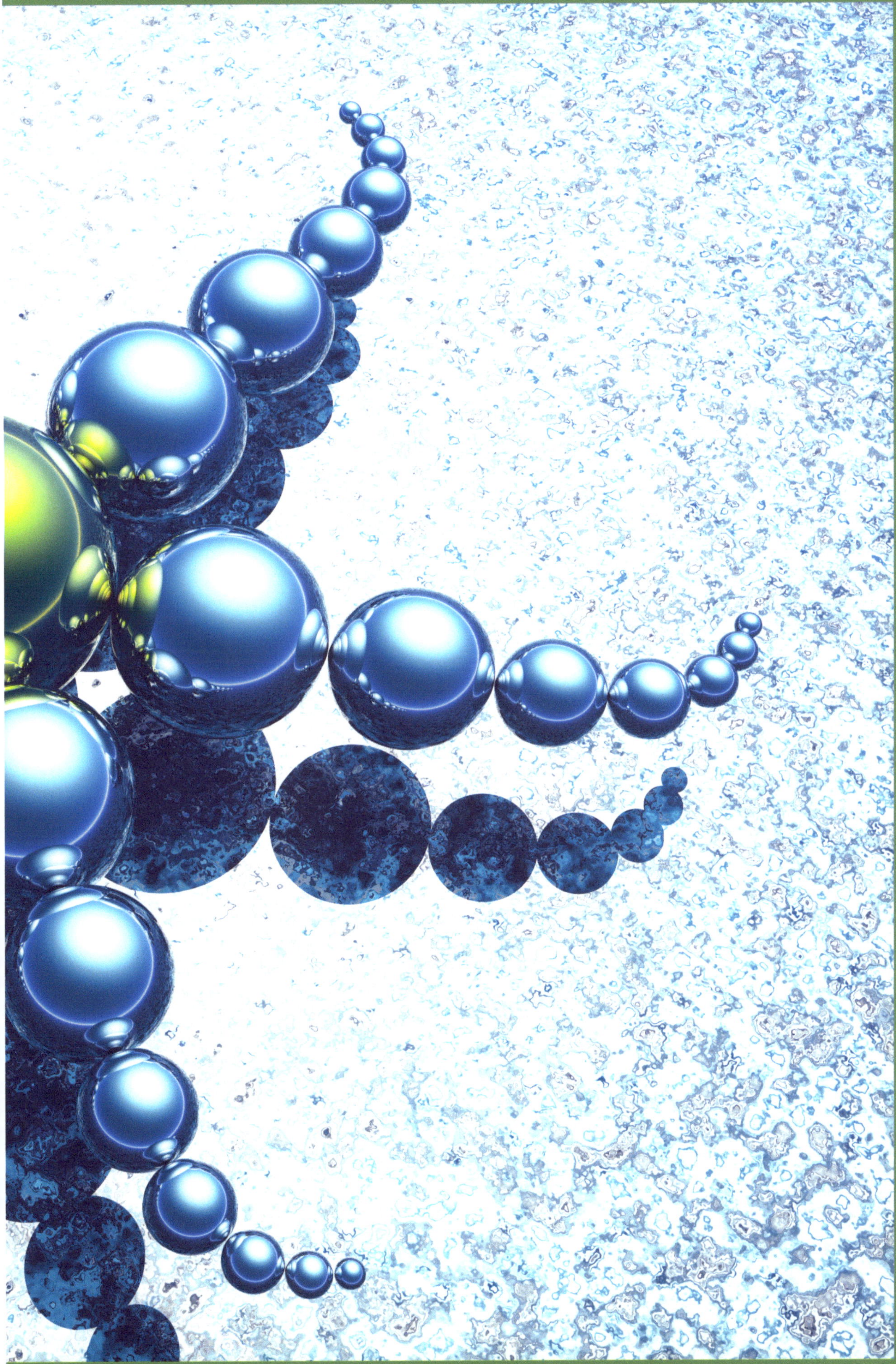

Previous pages:

BLUESTAR

EYE RING

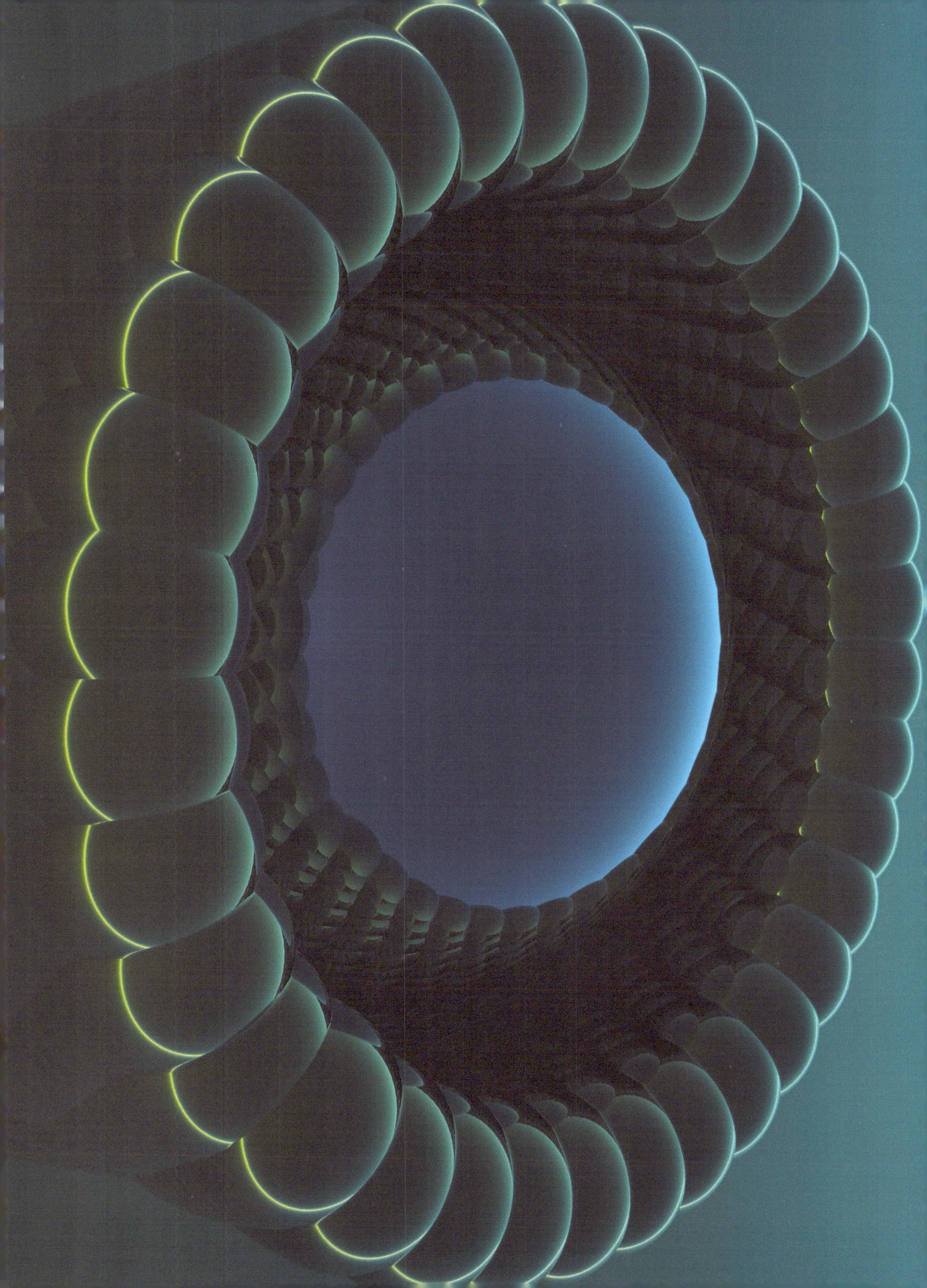

NEUROSIS

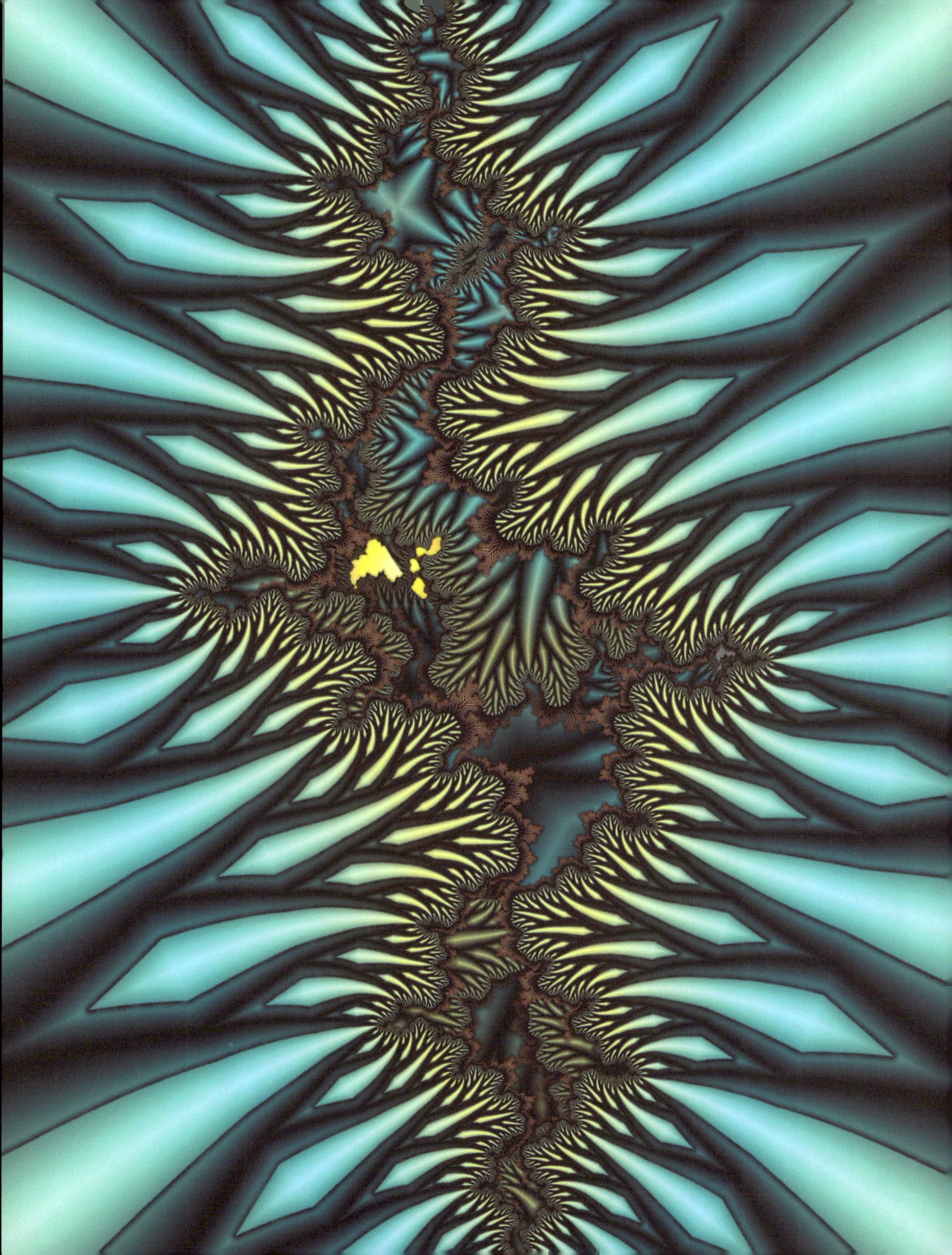

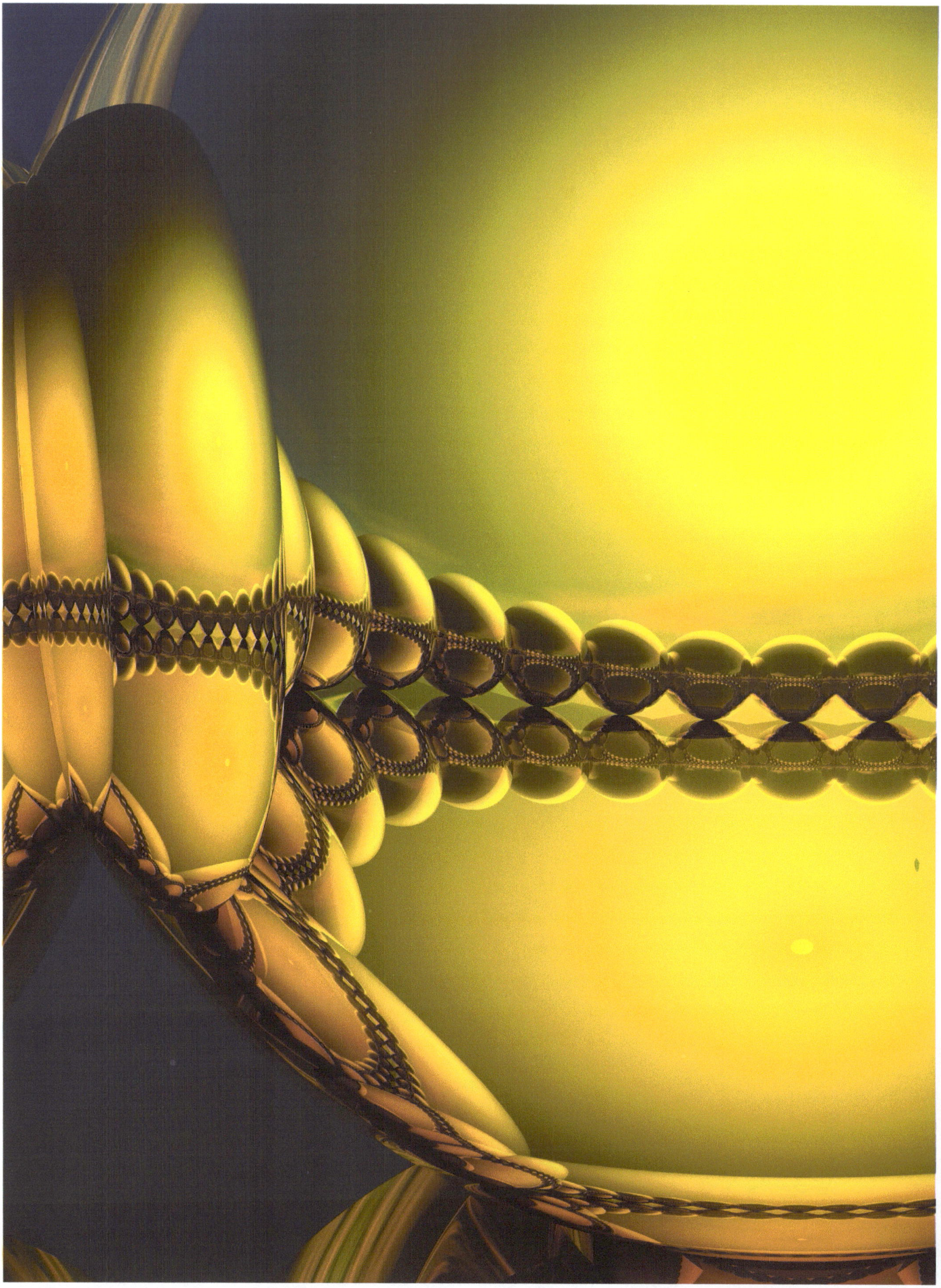

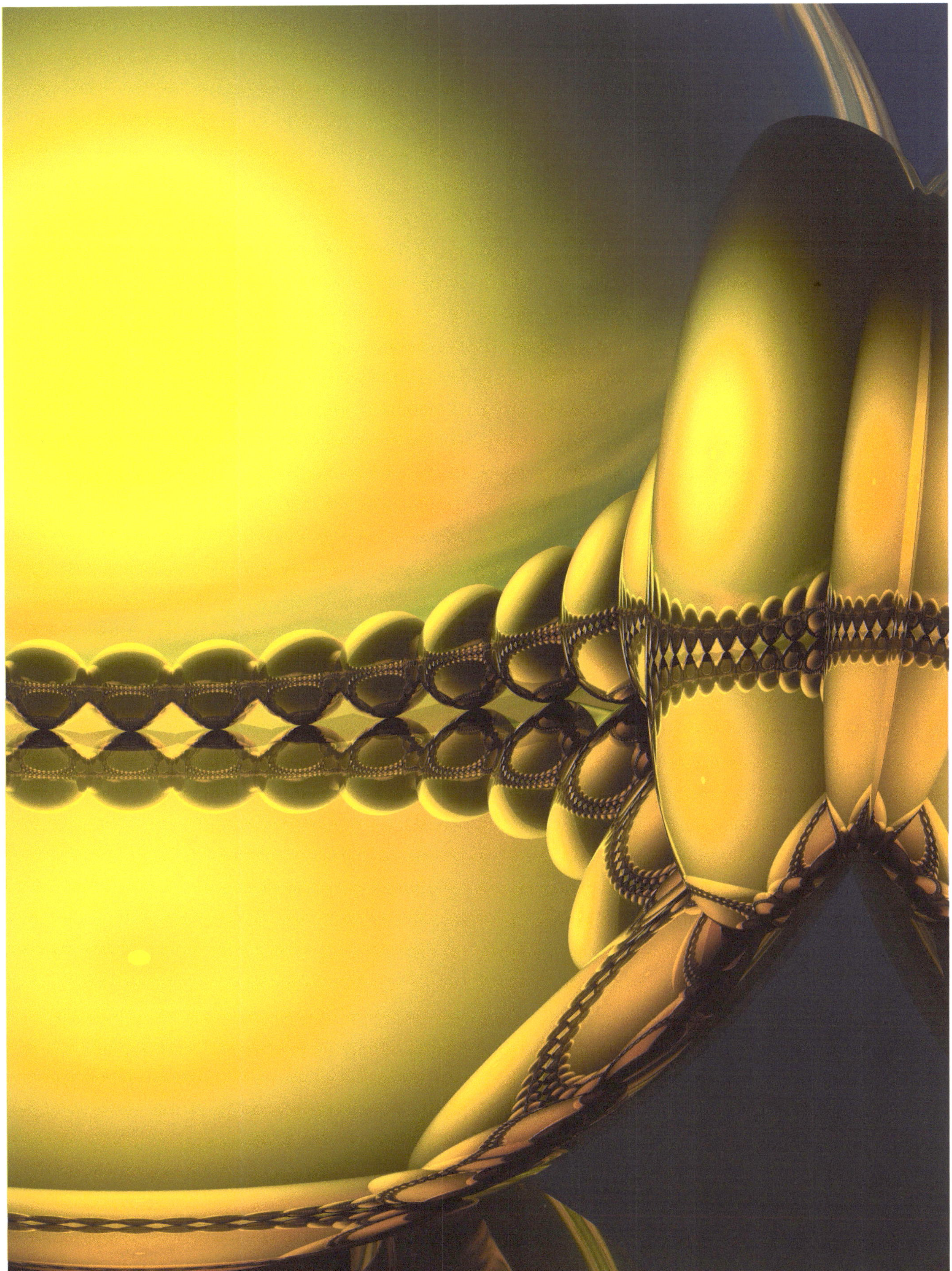

Previous pages:

GOLDEN ZIPPERS

IFNI

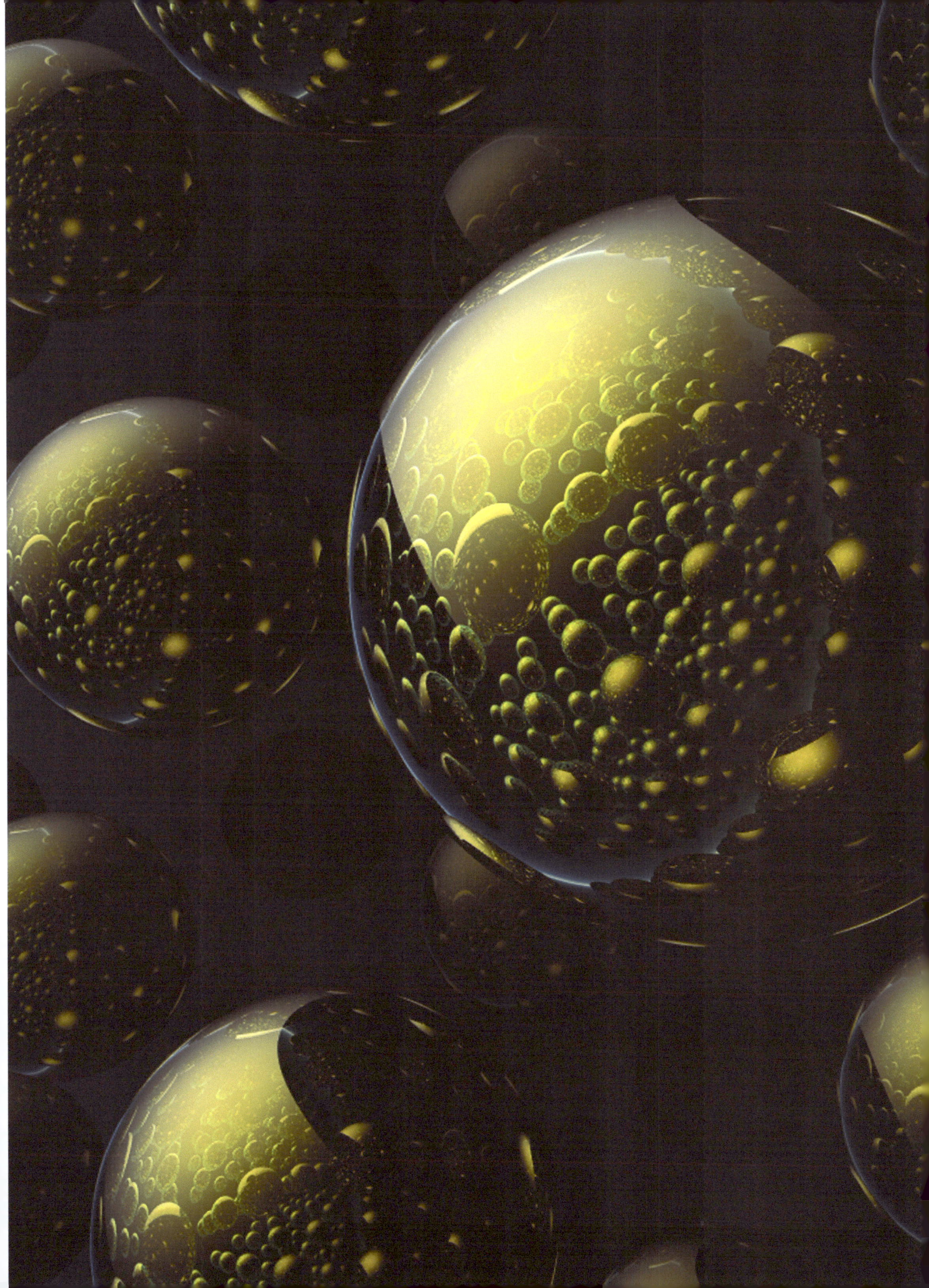

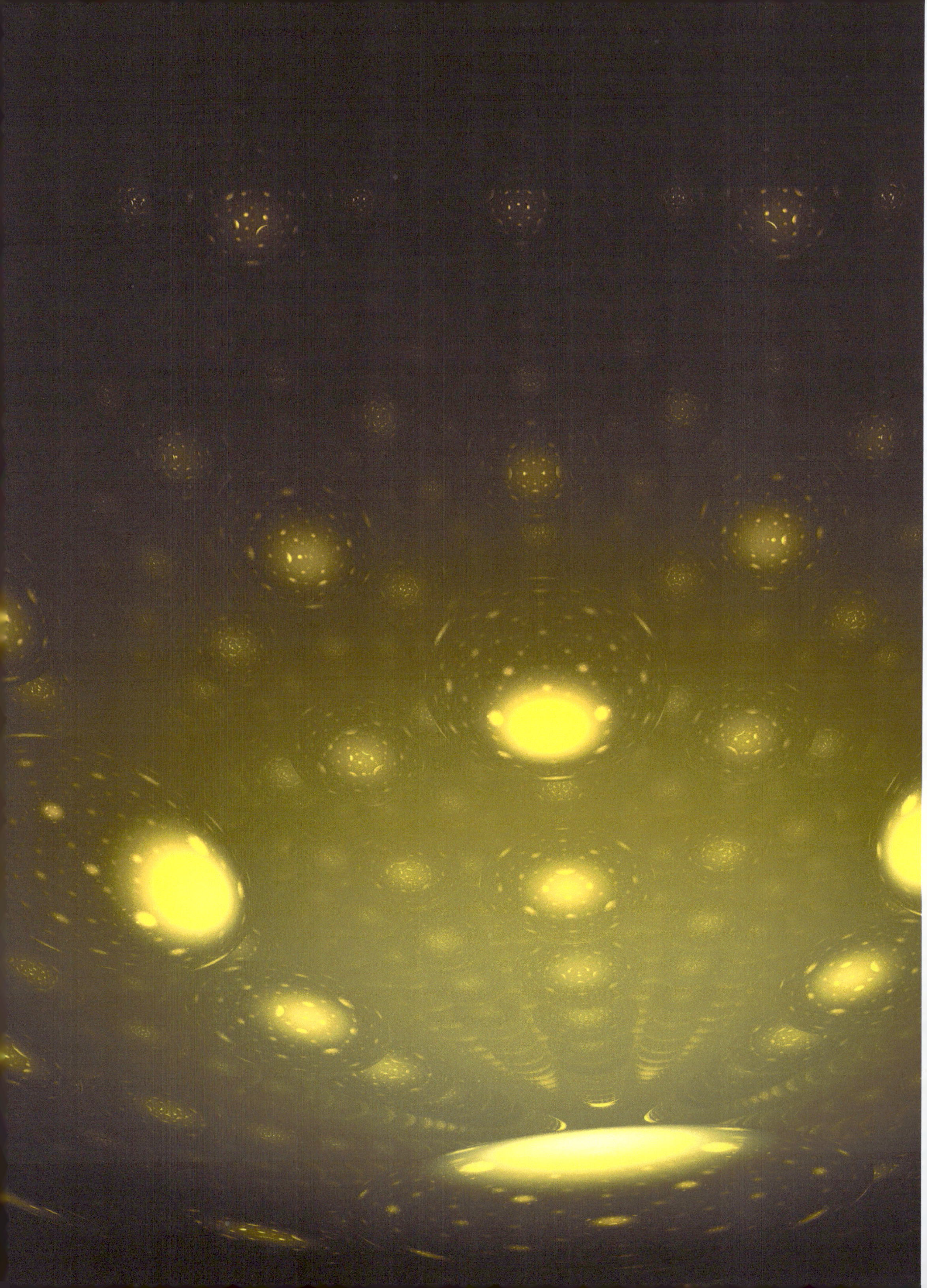

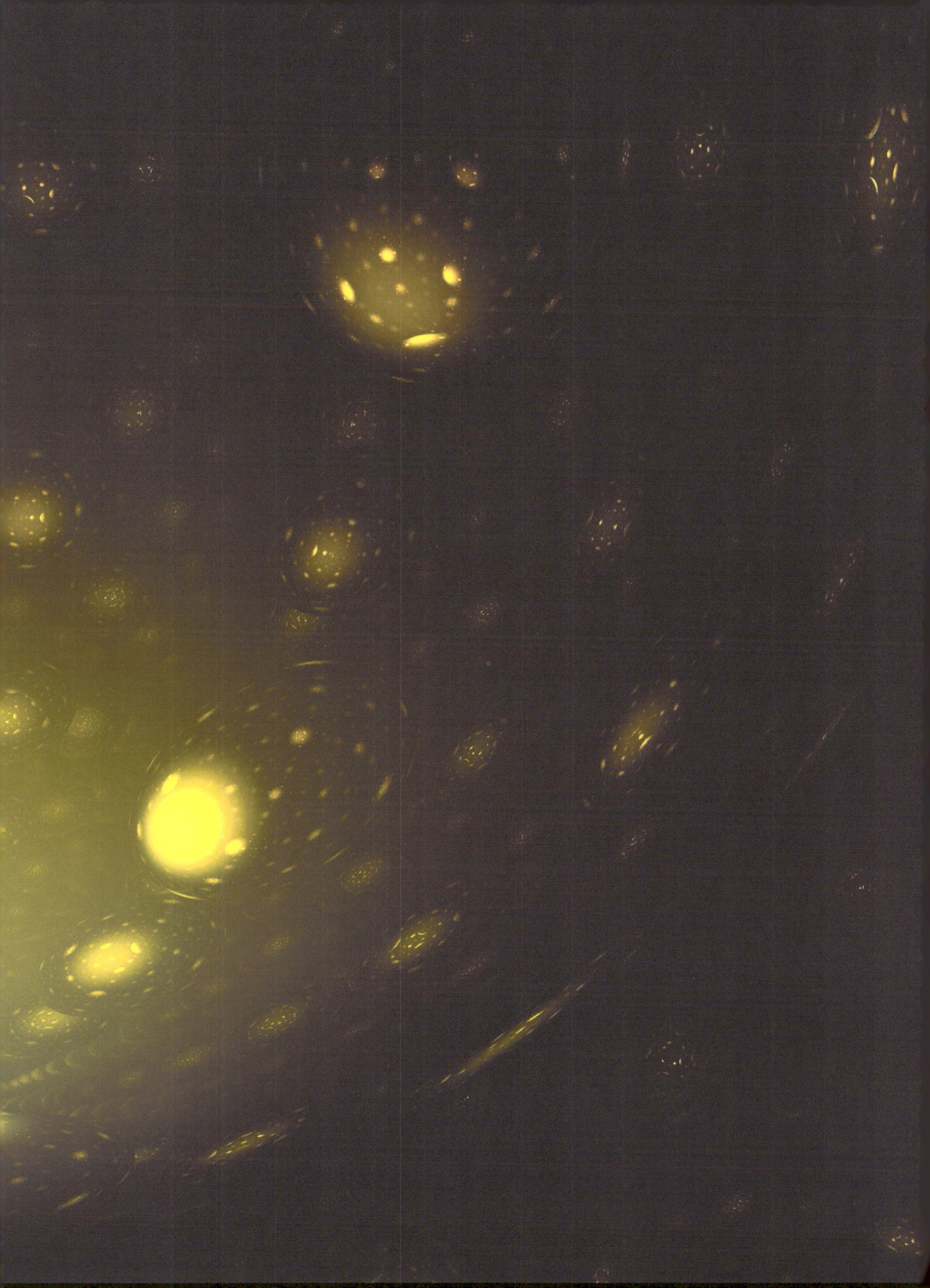

Previous pages:

IFNIFNI

GOLDEN DRAGON 2

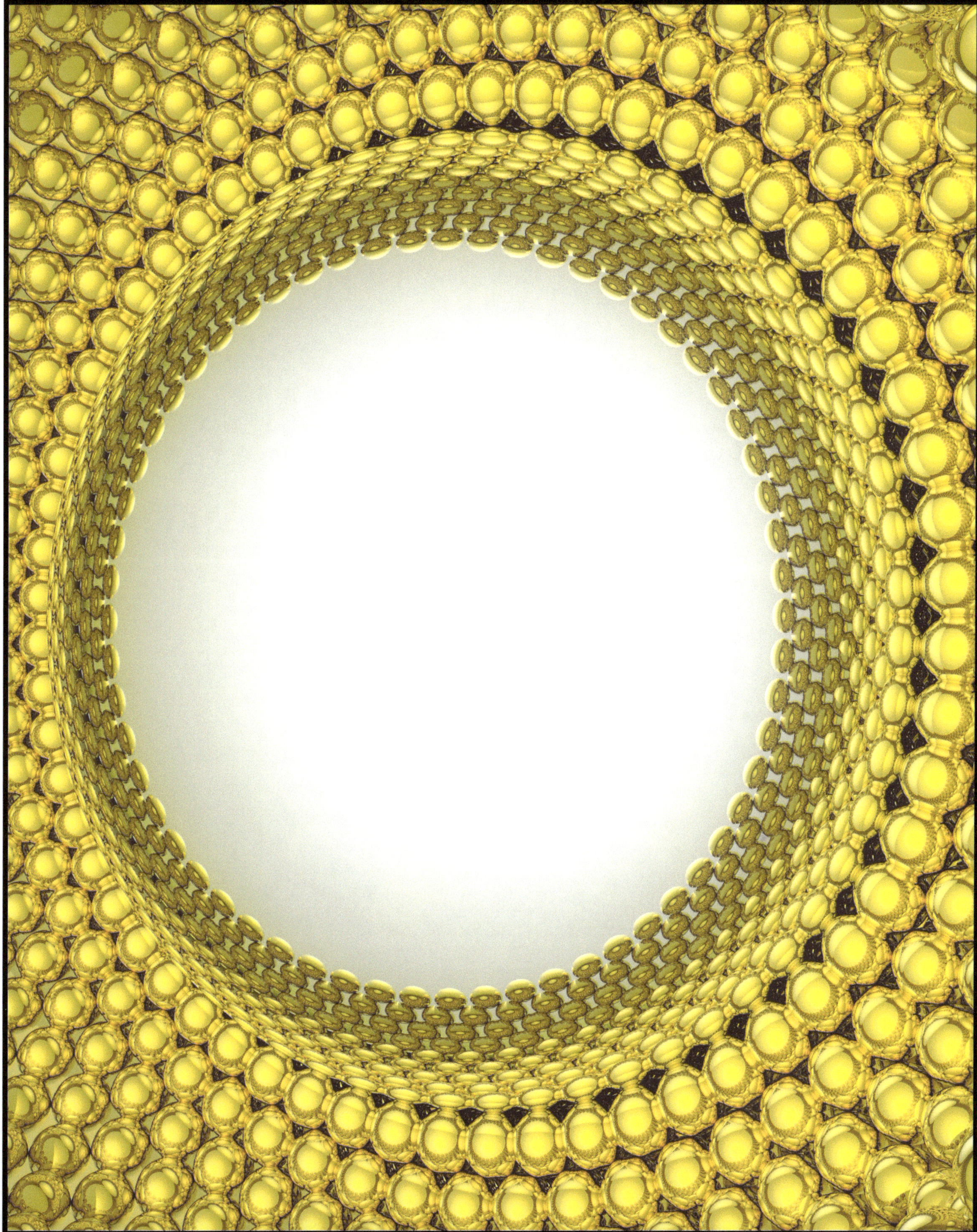

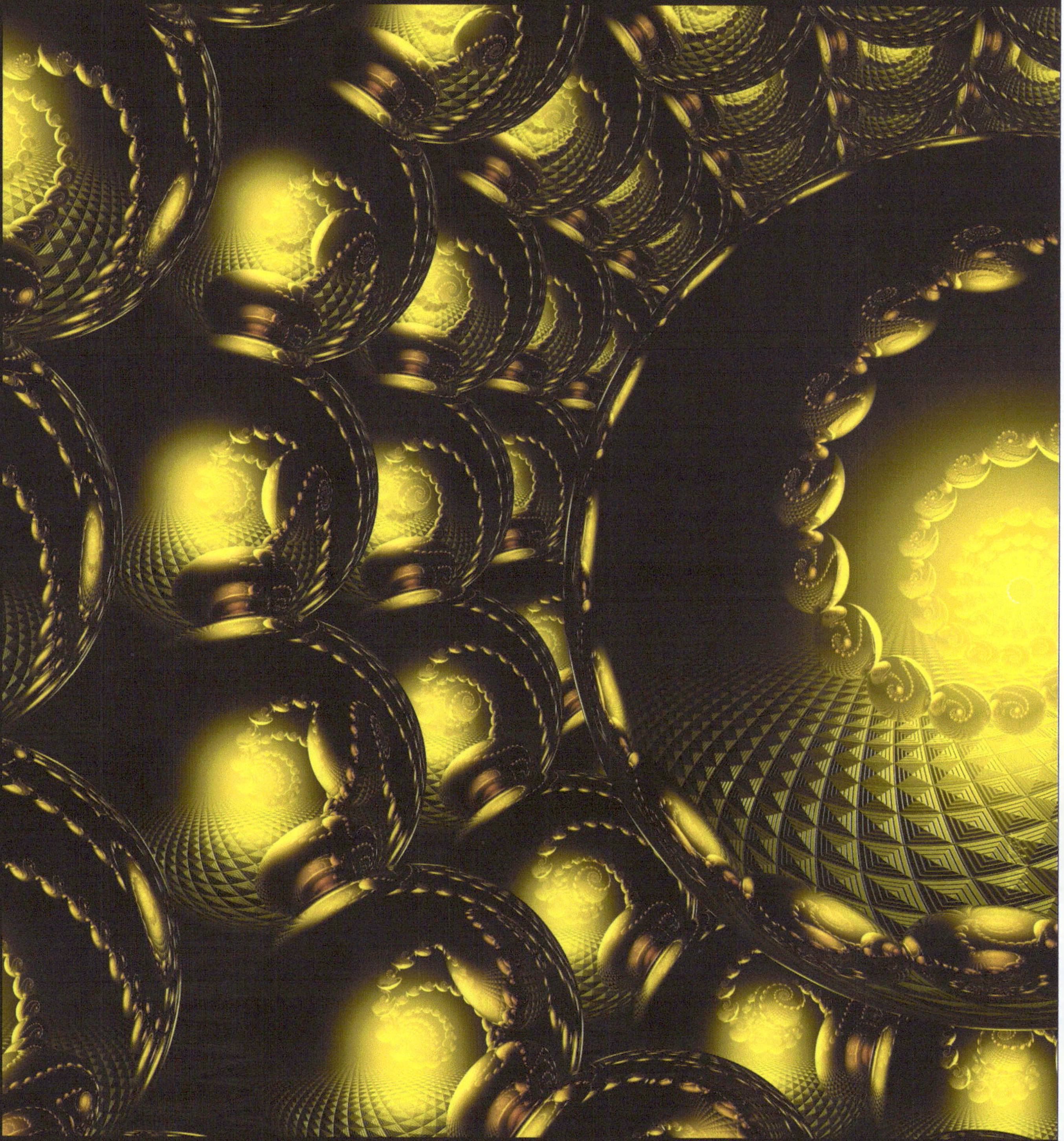

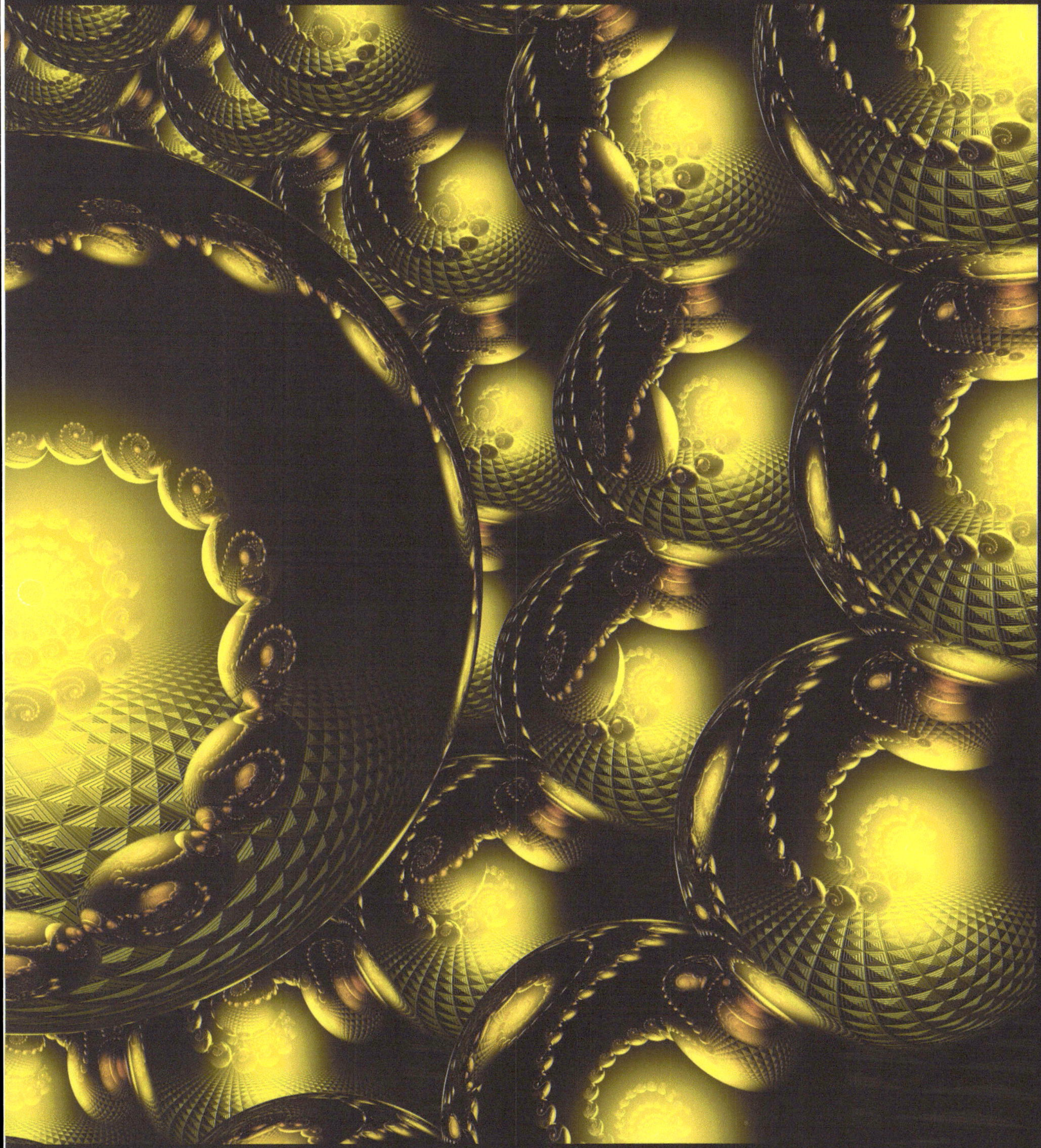

Previous pages:

NIGHTTRIPPER

DIGI-LION

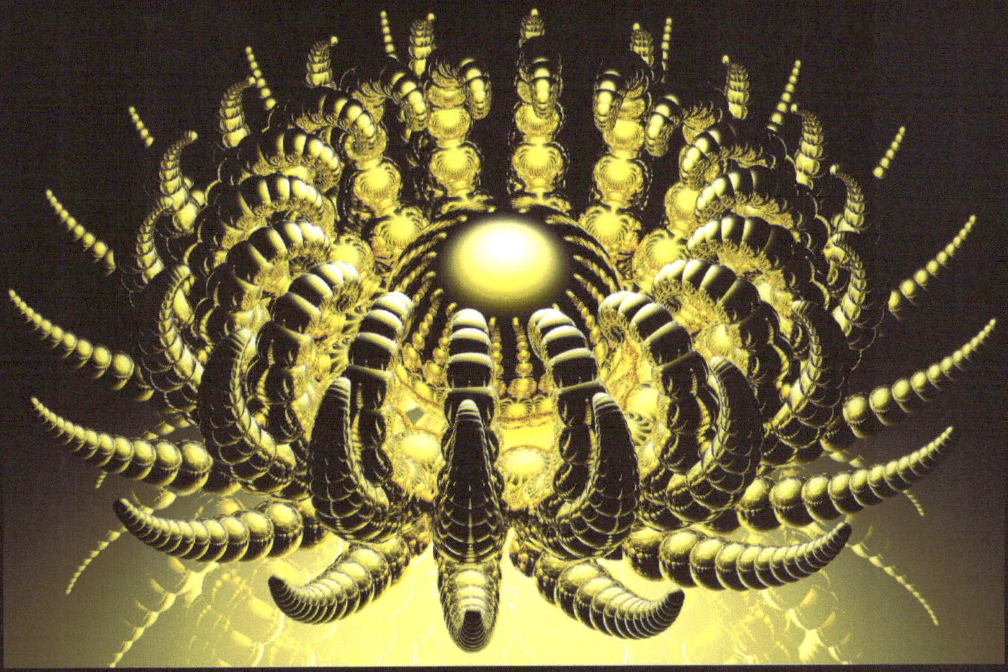

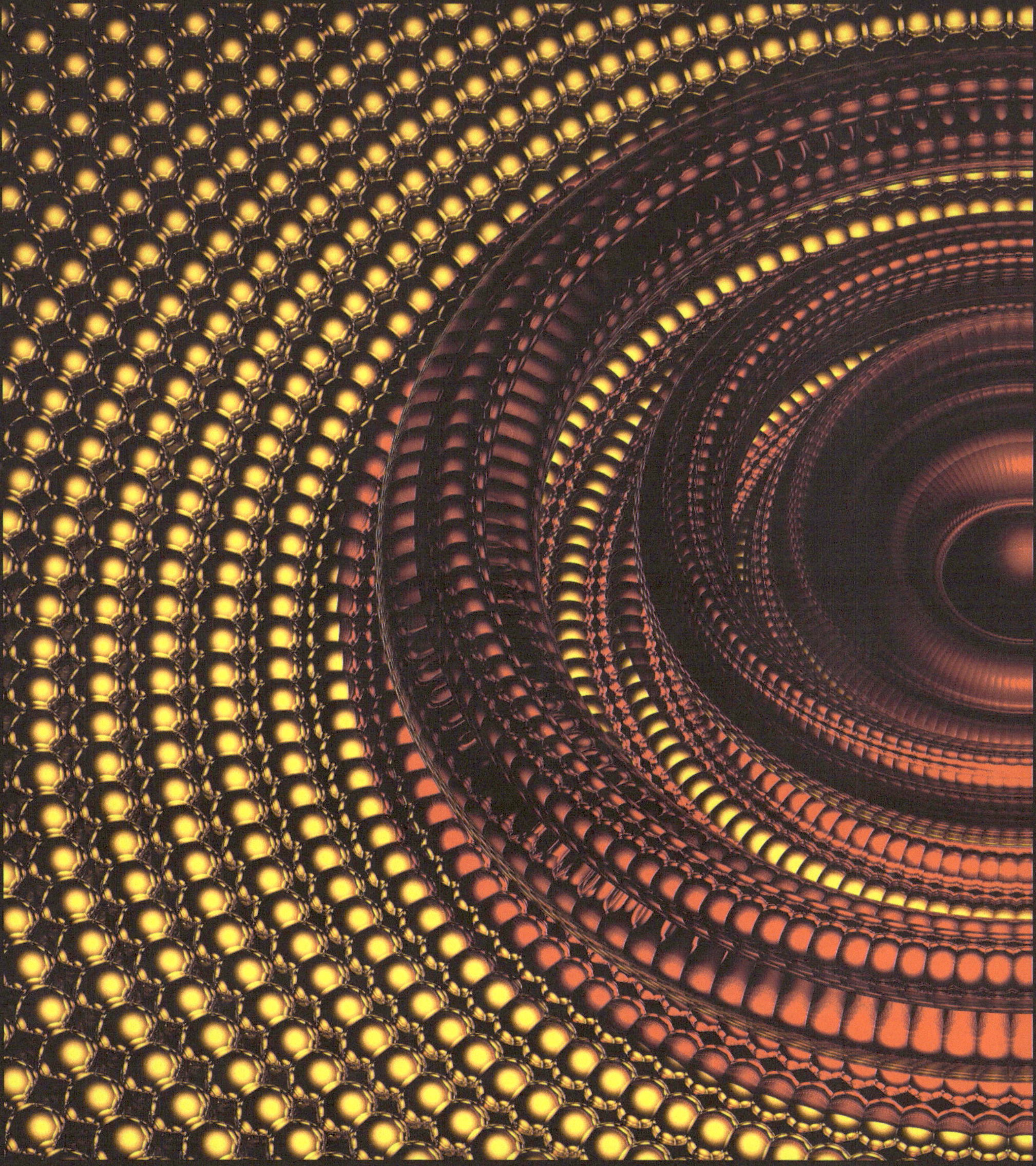

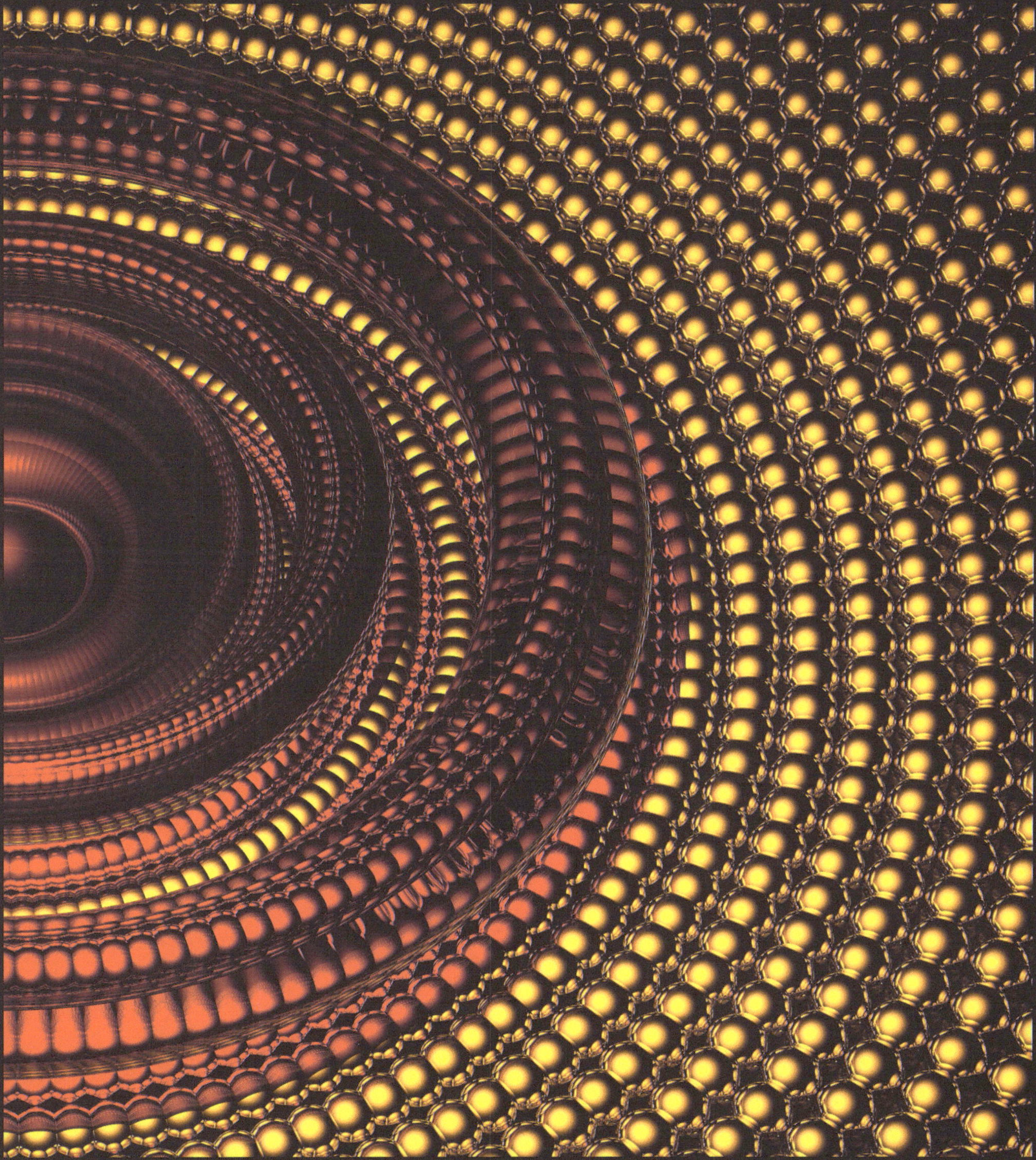

Previous pages:

STARGAZER

DARKFIRE

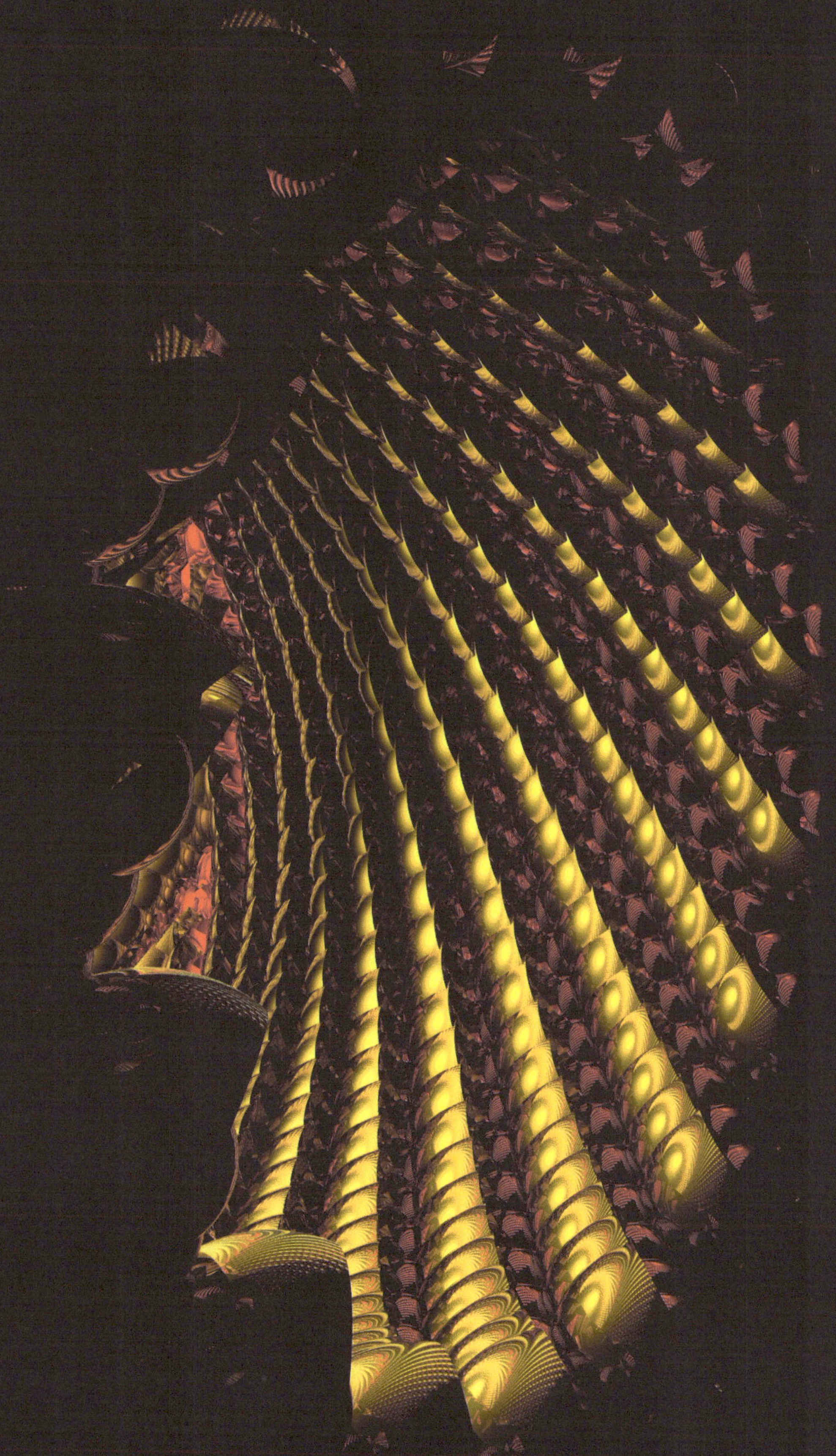

Previous pages:

MOONLIGHT

FRACTAL HEART

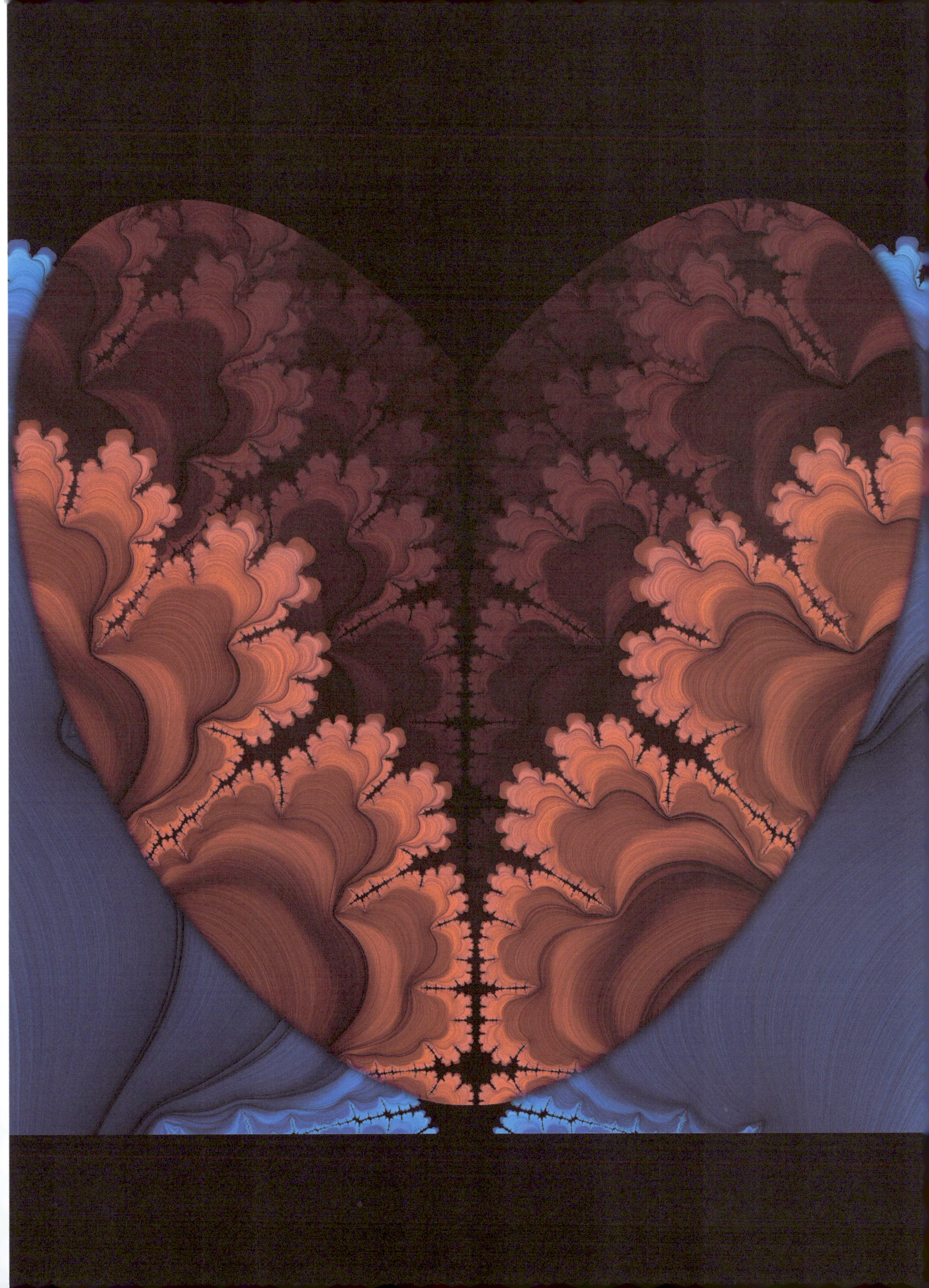

BLUESCALE

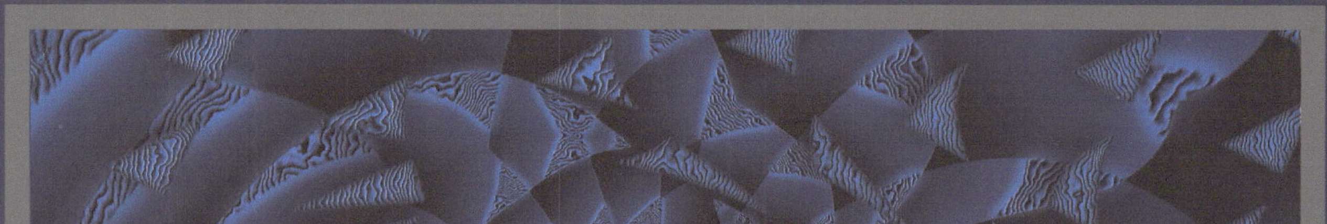

IONBLANKET

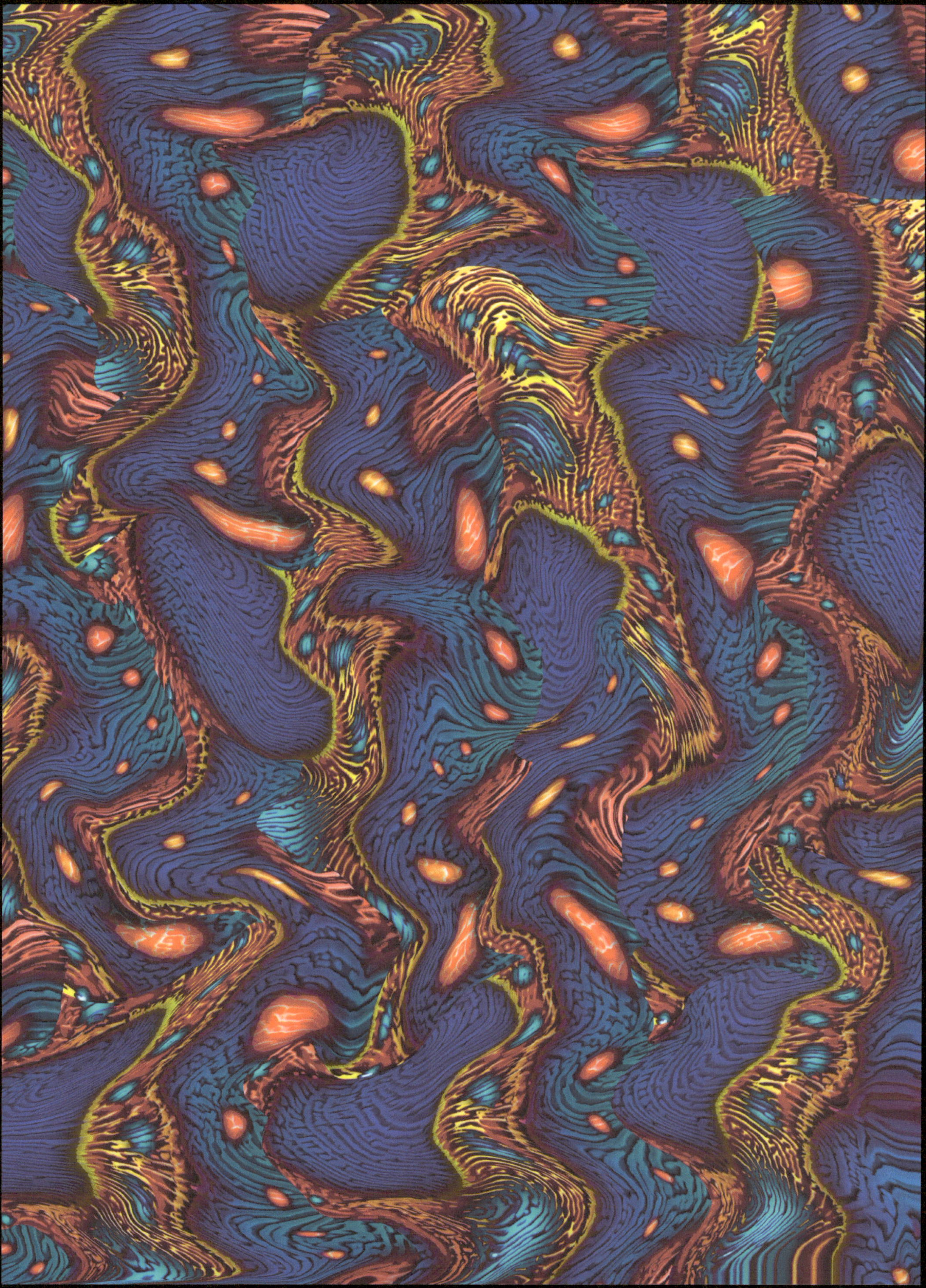

IQL

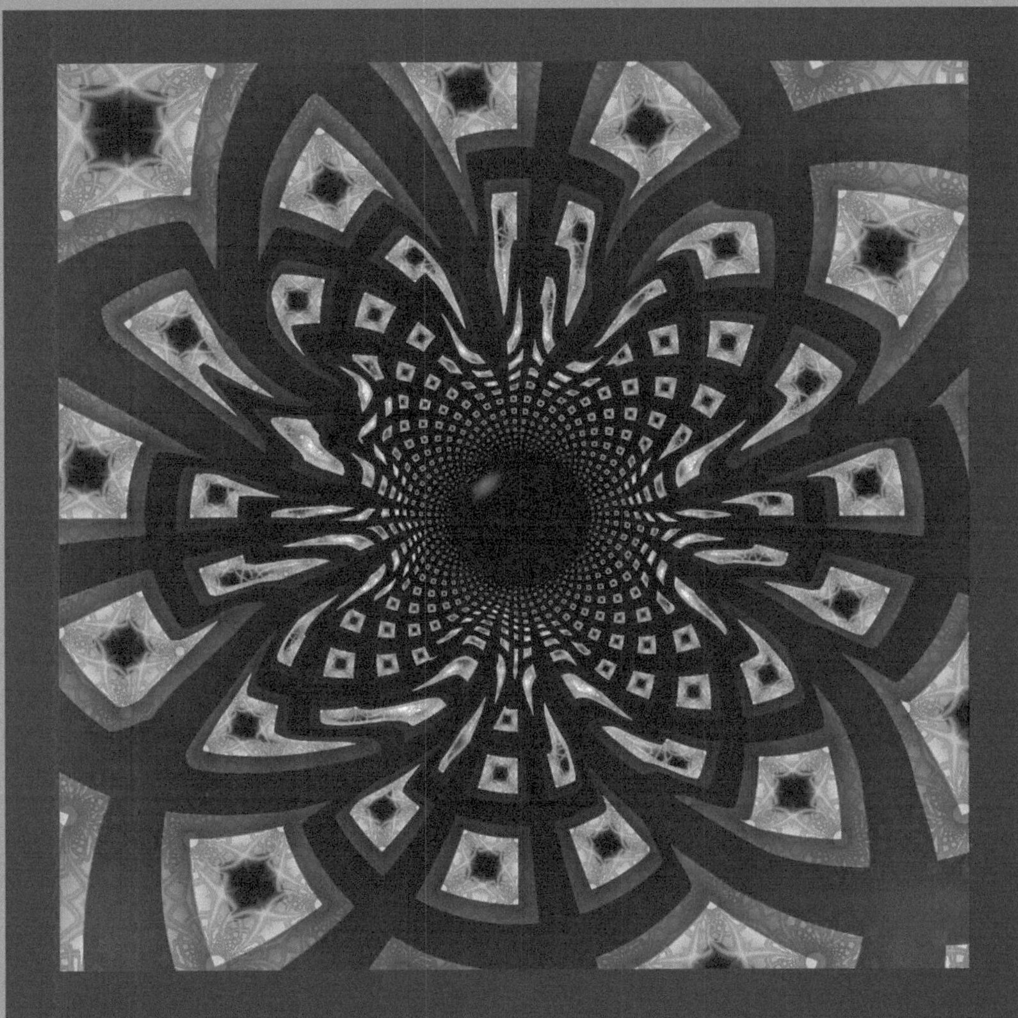

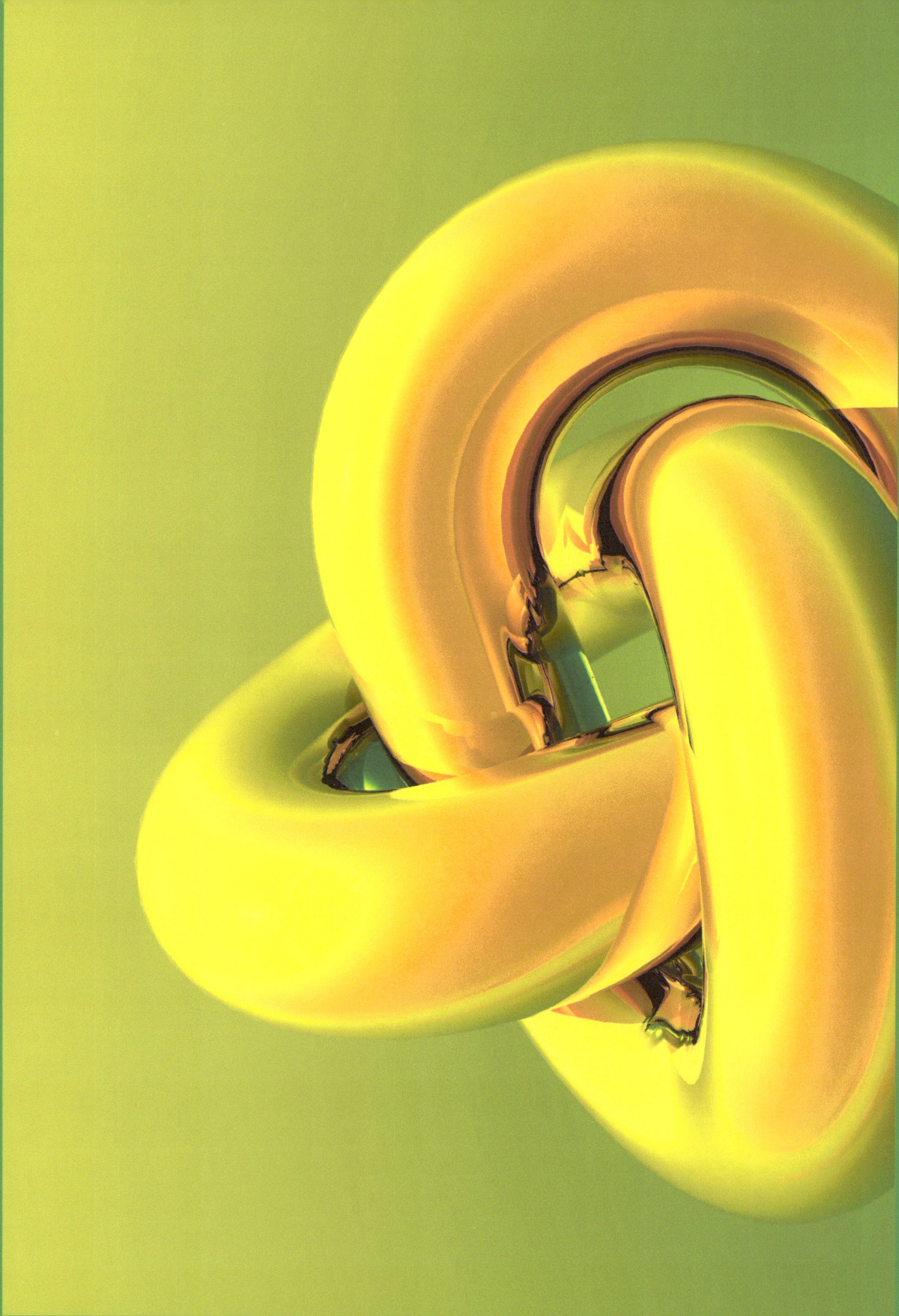

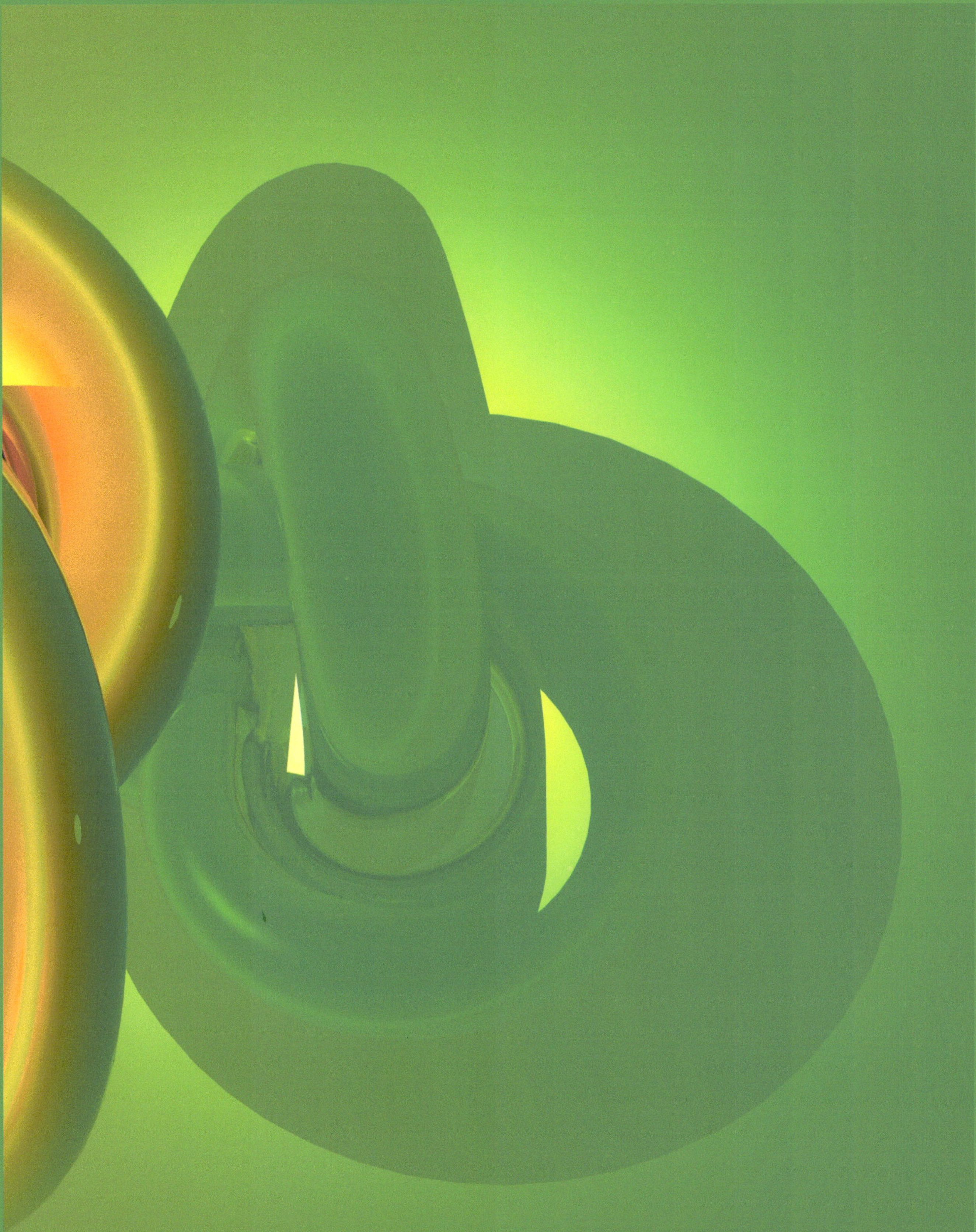

Previous pages:

KNOT

PETRIDISH

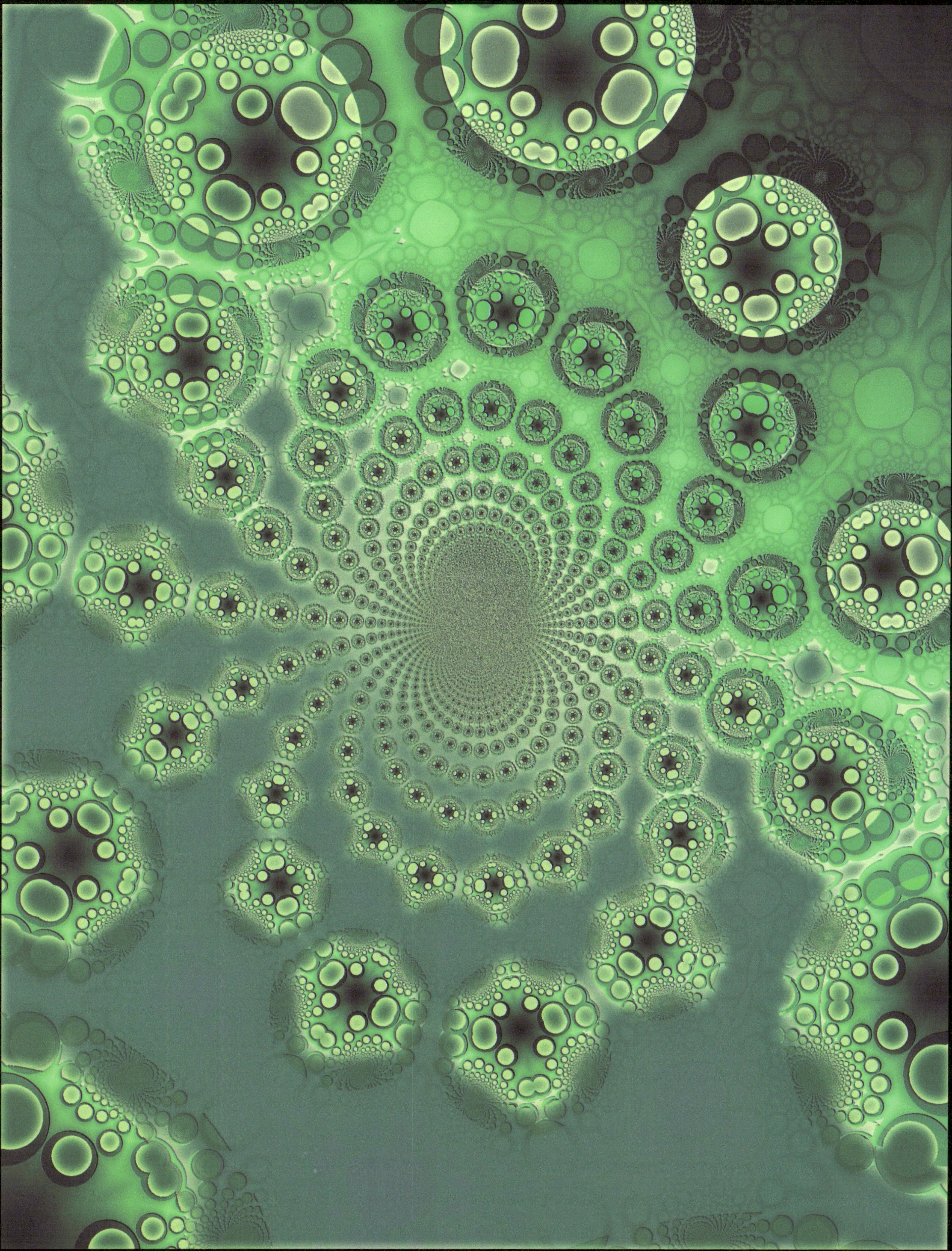

EARTHCUP SERIES

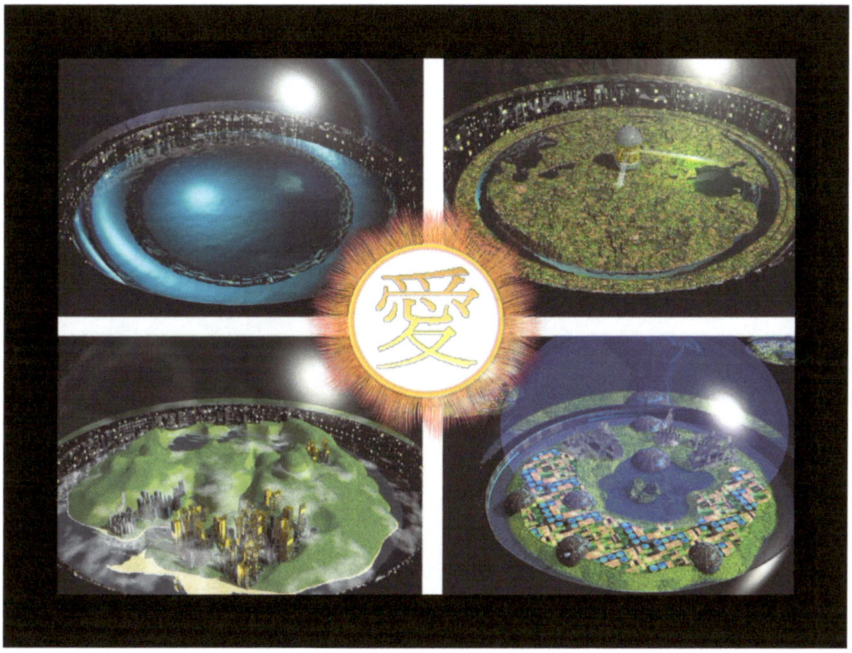
愛

BUTTERFLYWEAVE 1 & 2

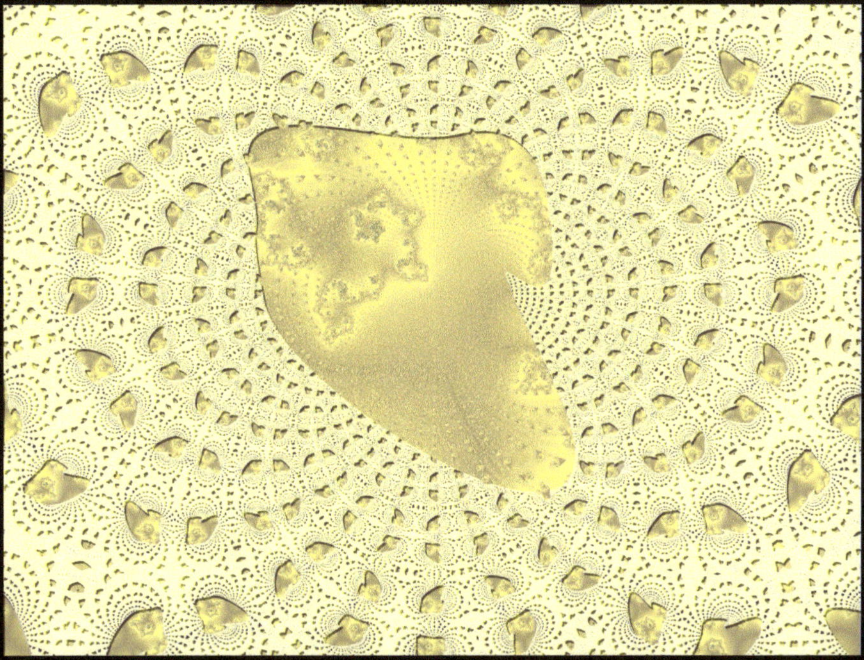

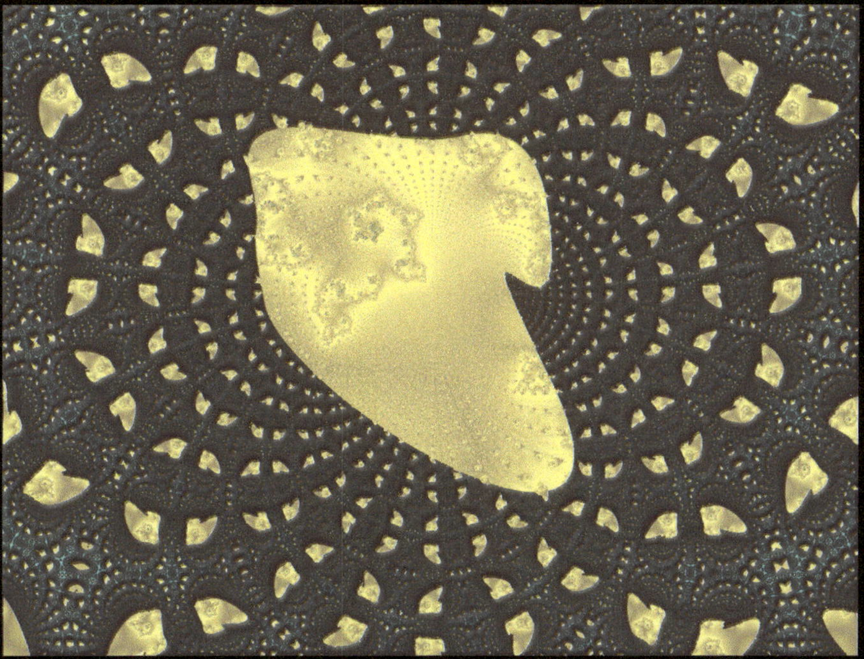

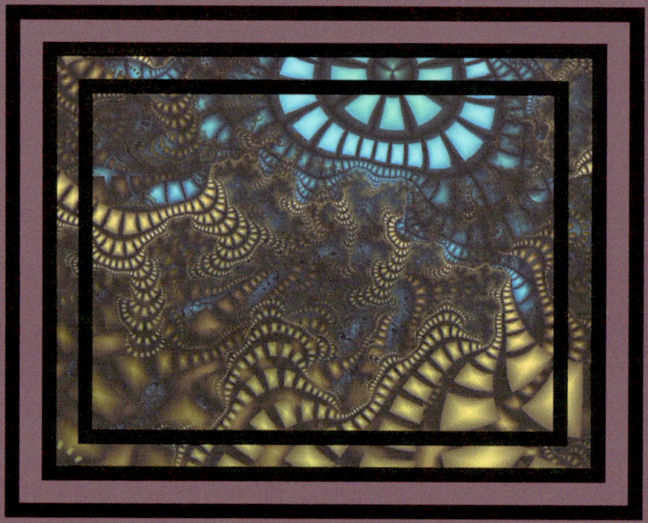

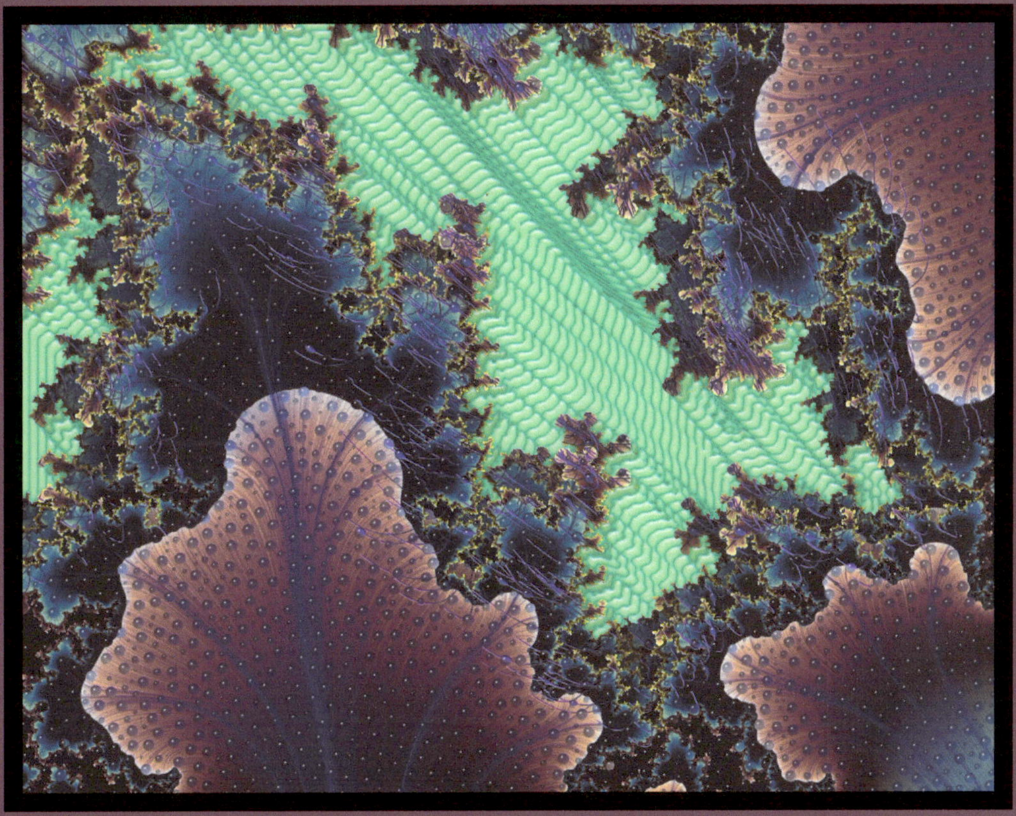

Previous pages:

OILY COLLECTION

IRRIDIUM

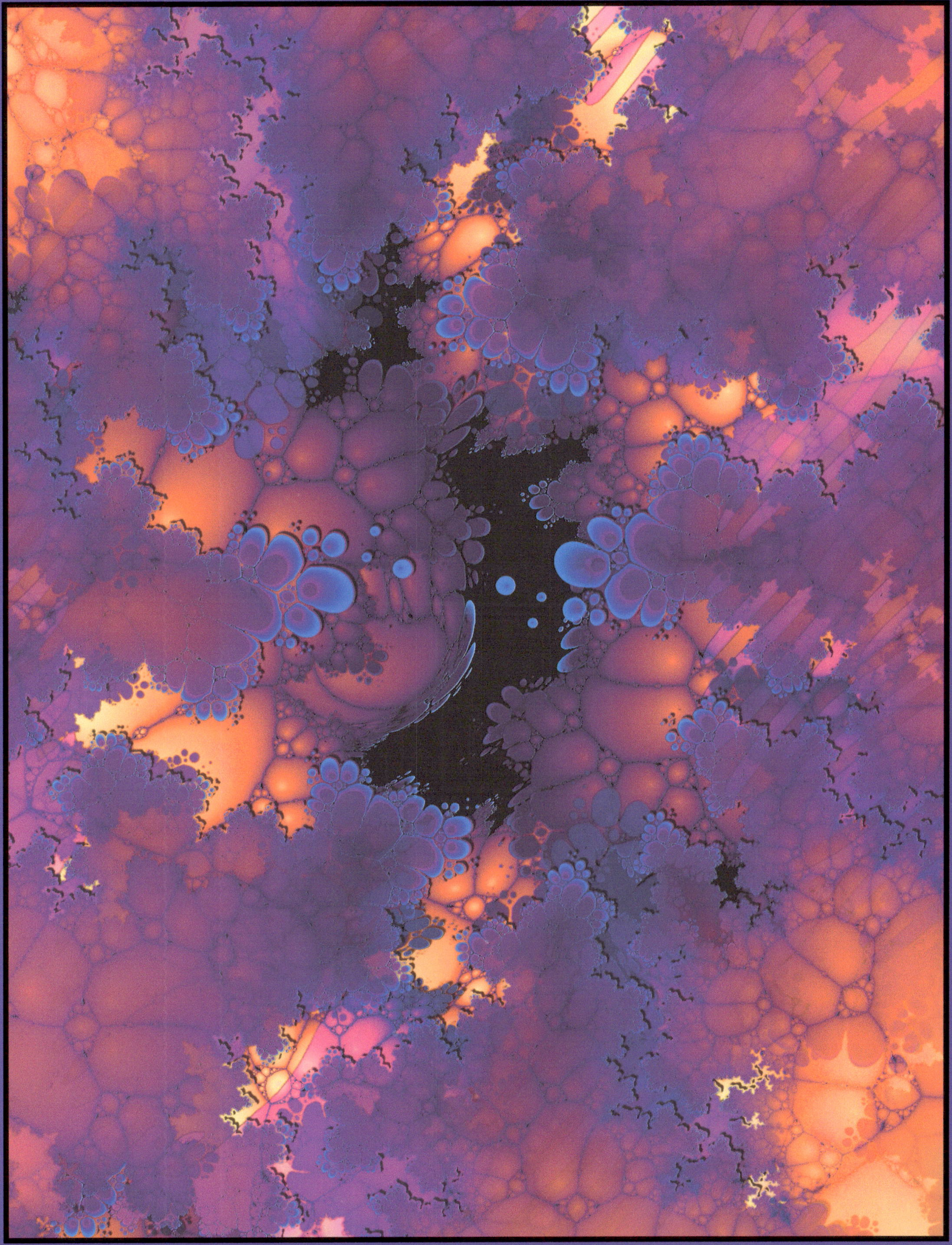

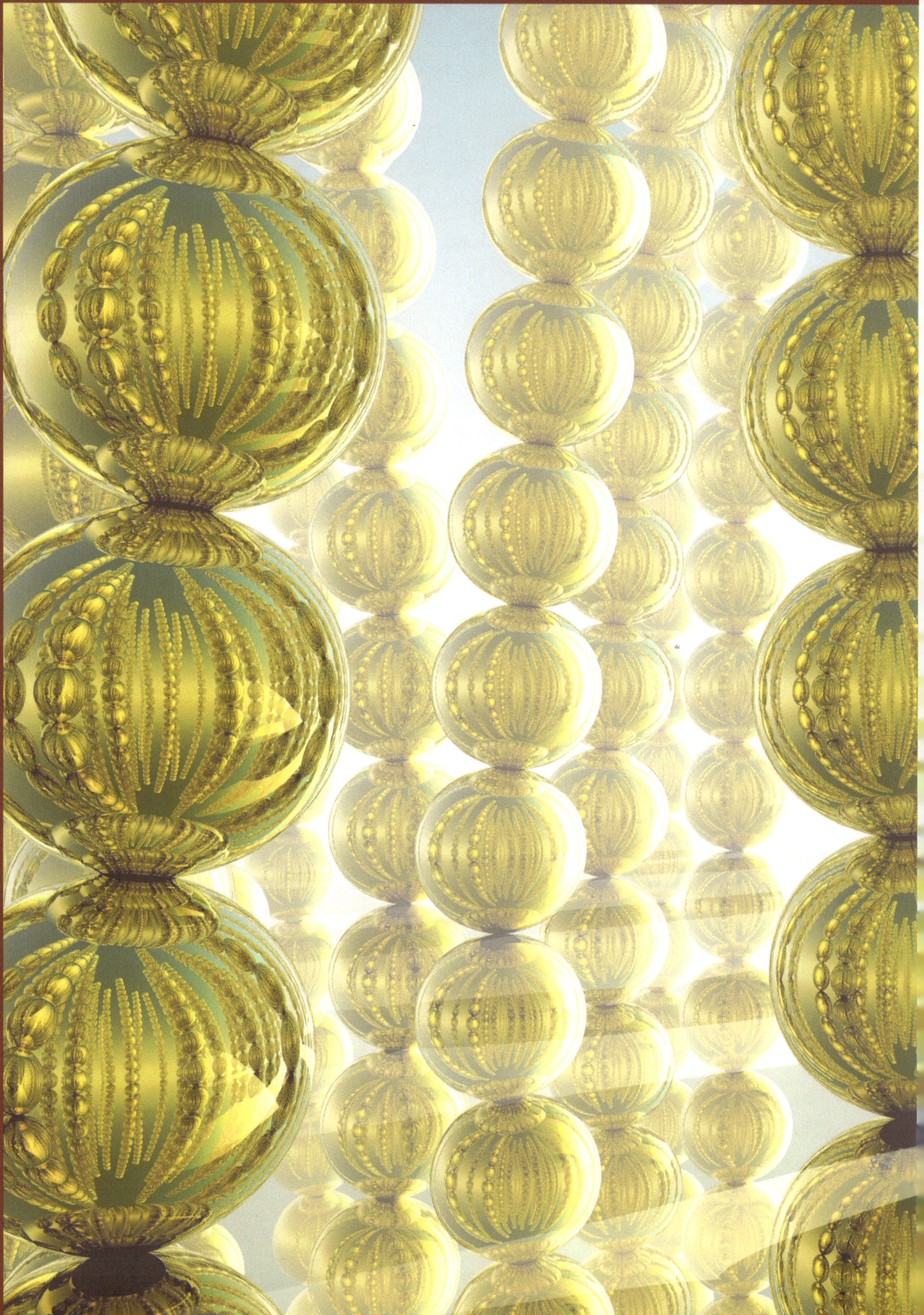

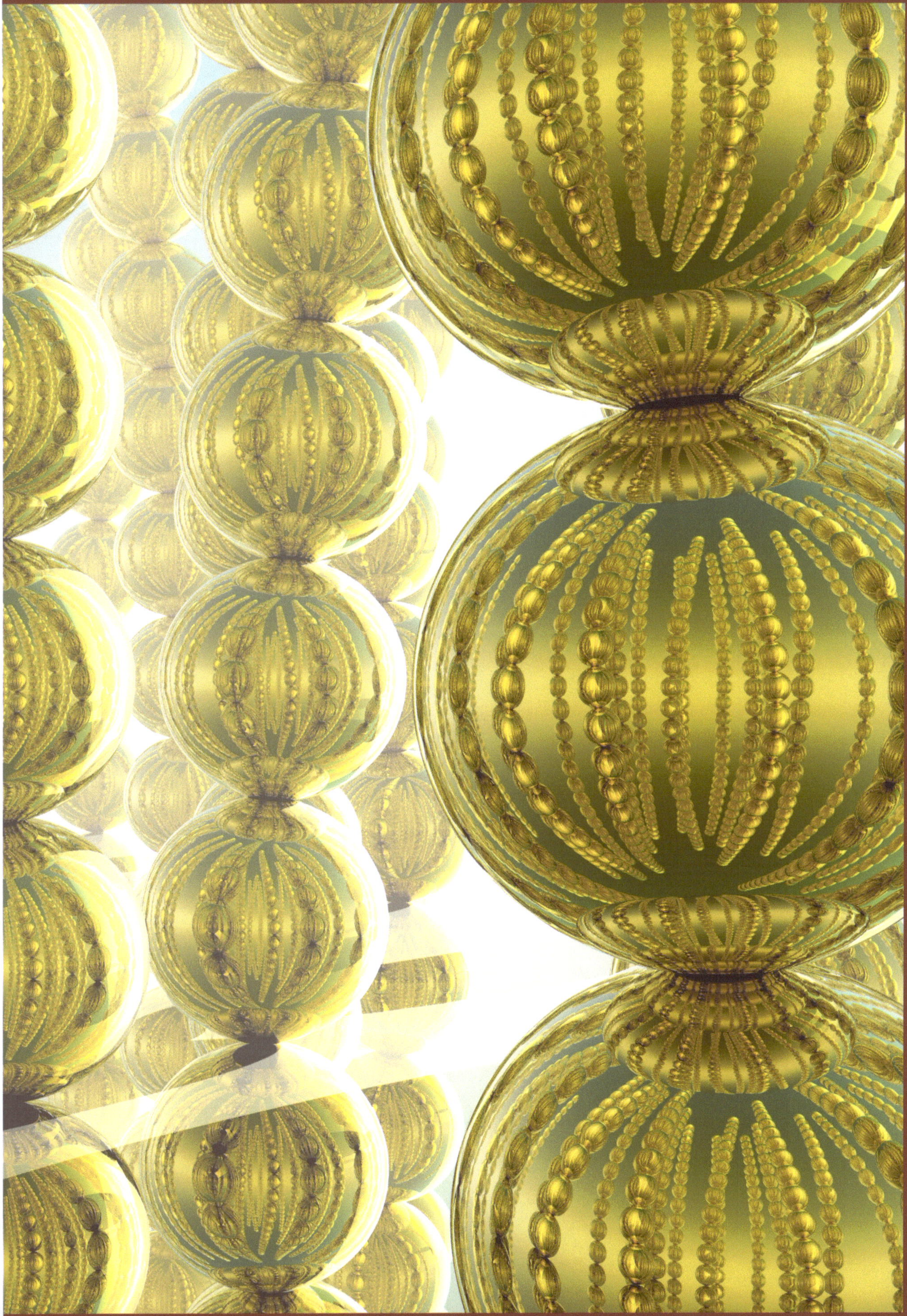

Previous pages:

XQWIZIT

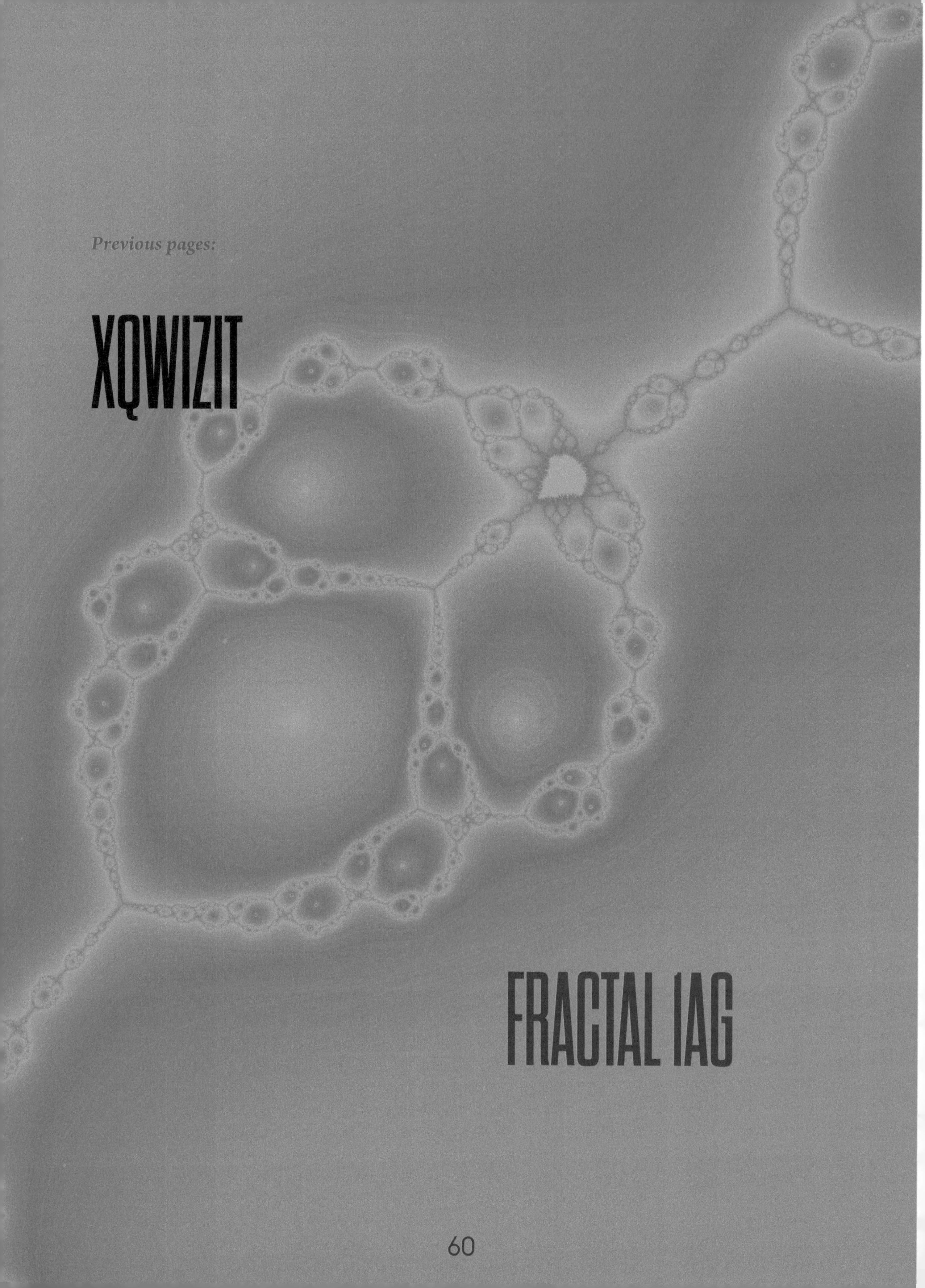

FRACTAL IAG

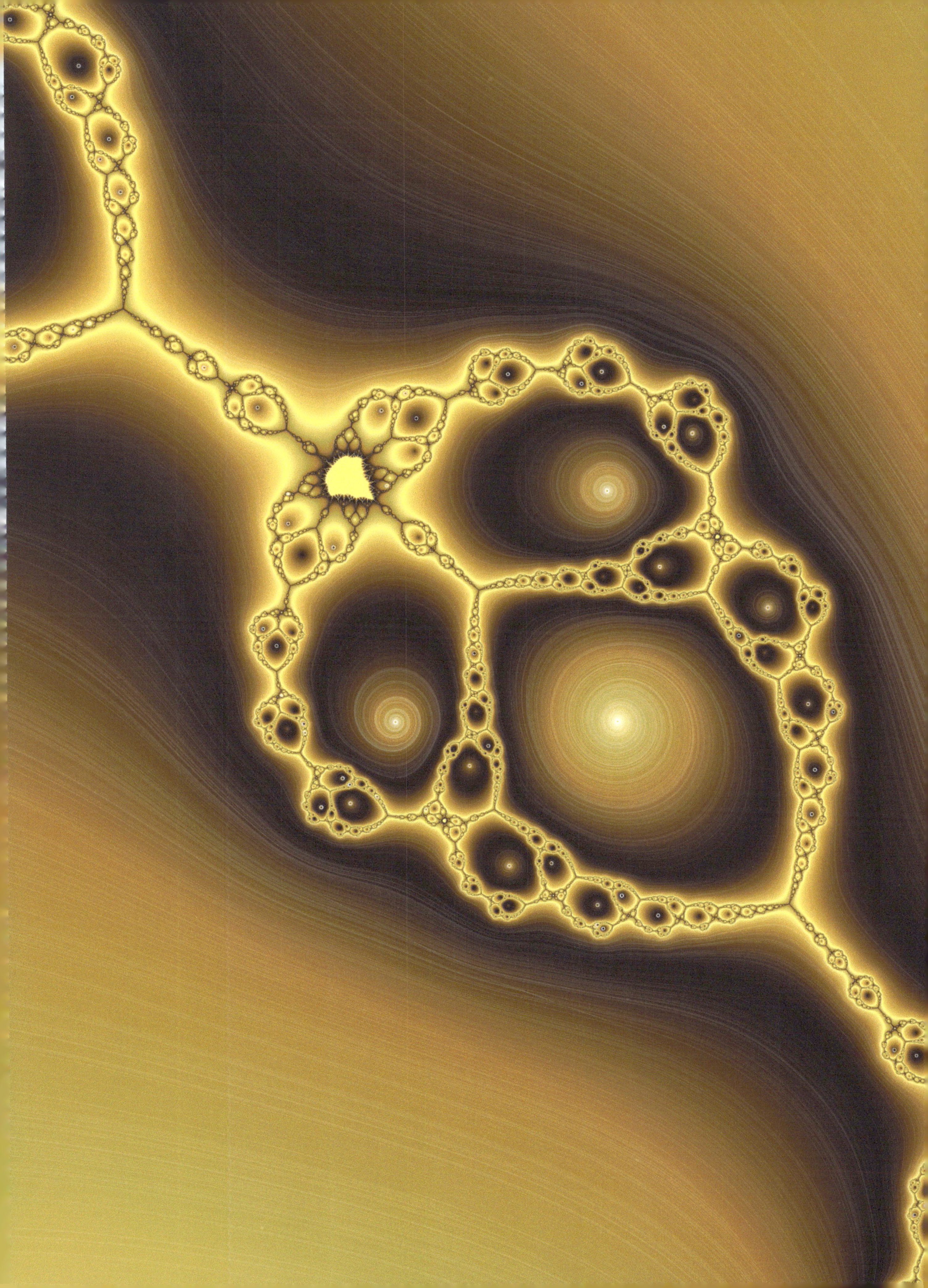

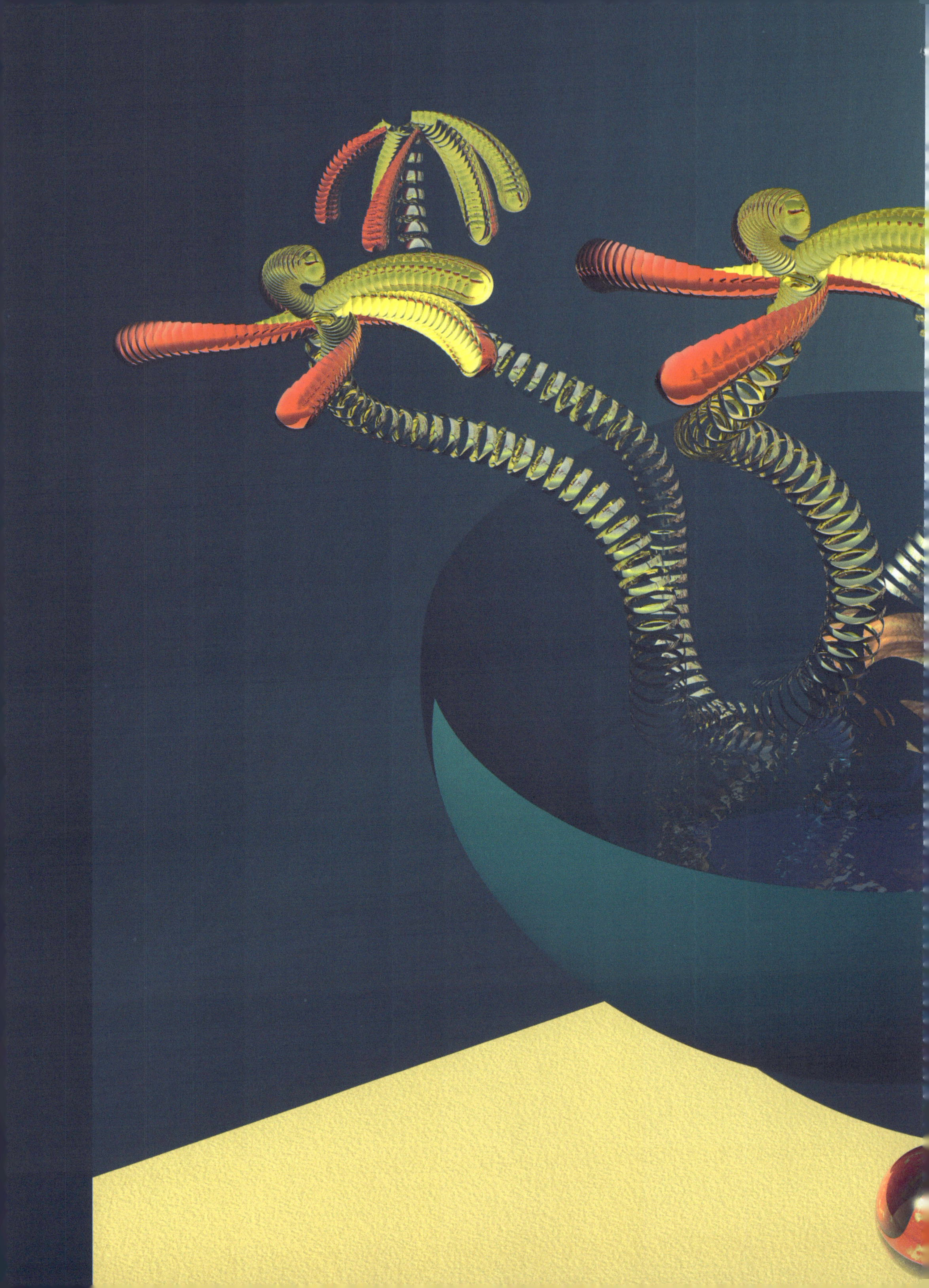

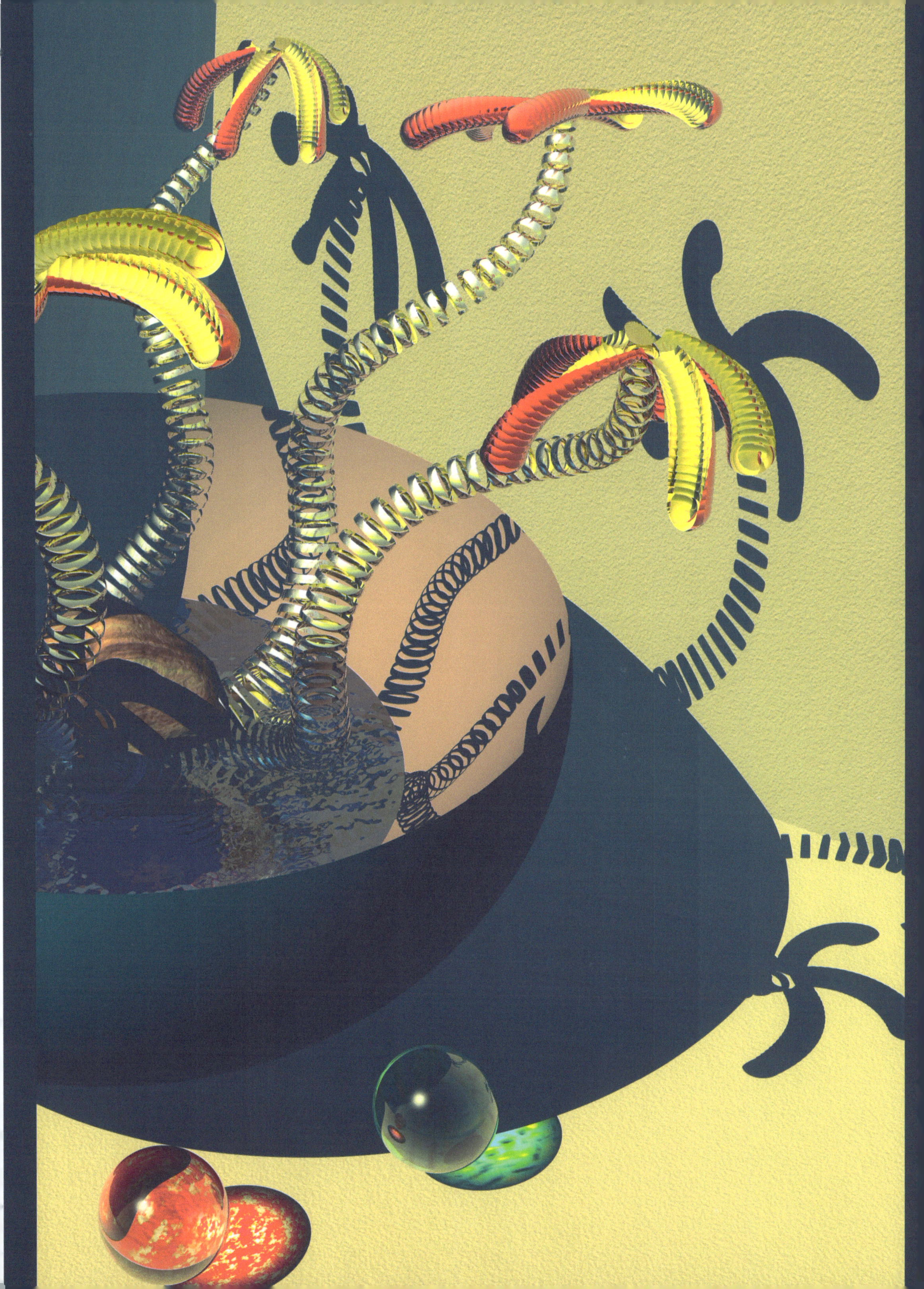

Previous pages:

CYBERBOQUET

FRACTIQUE

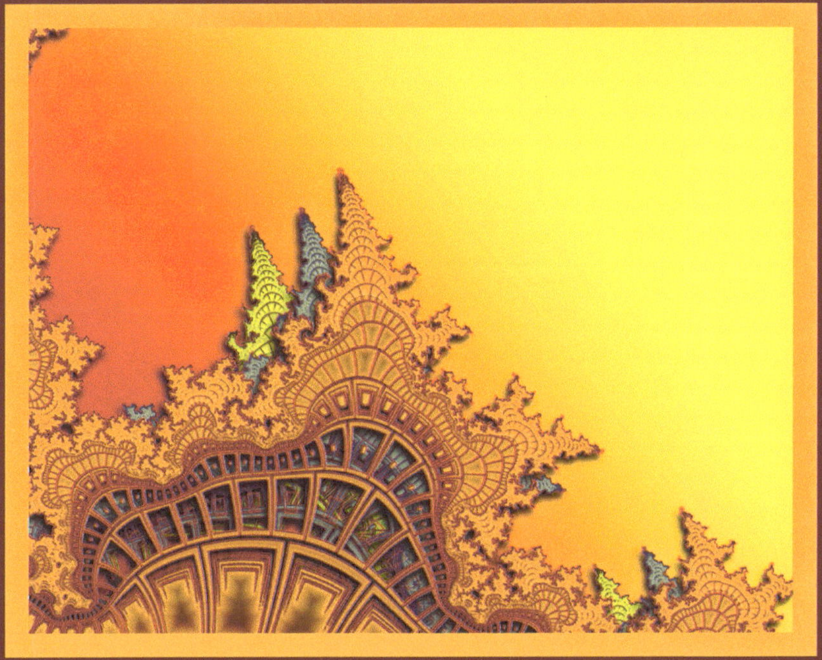

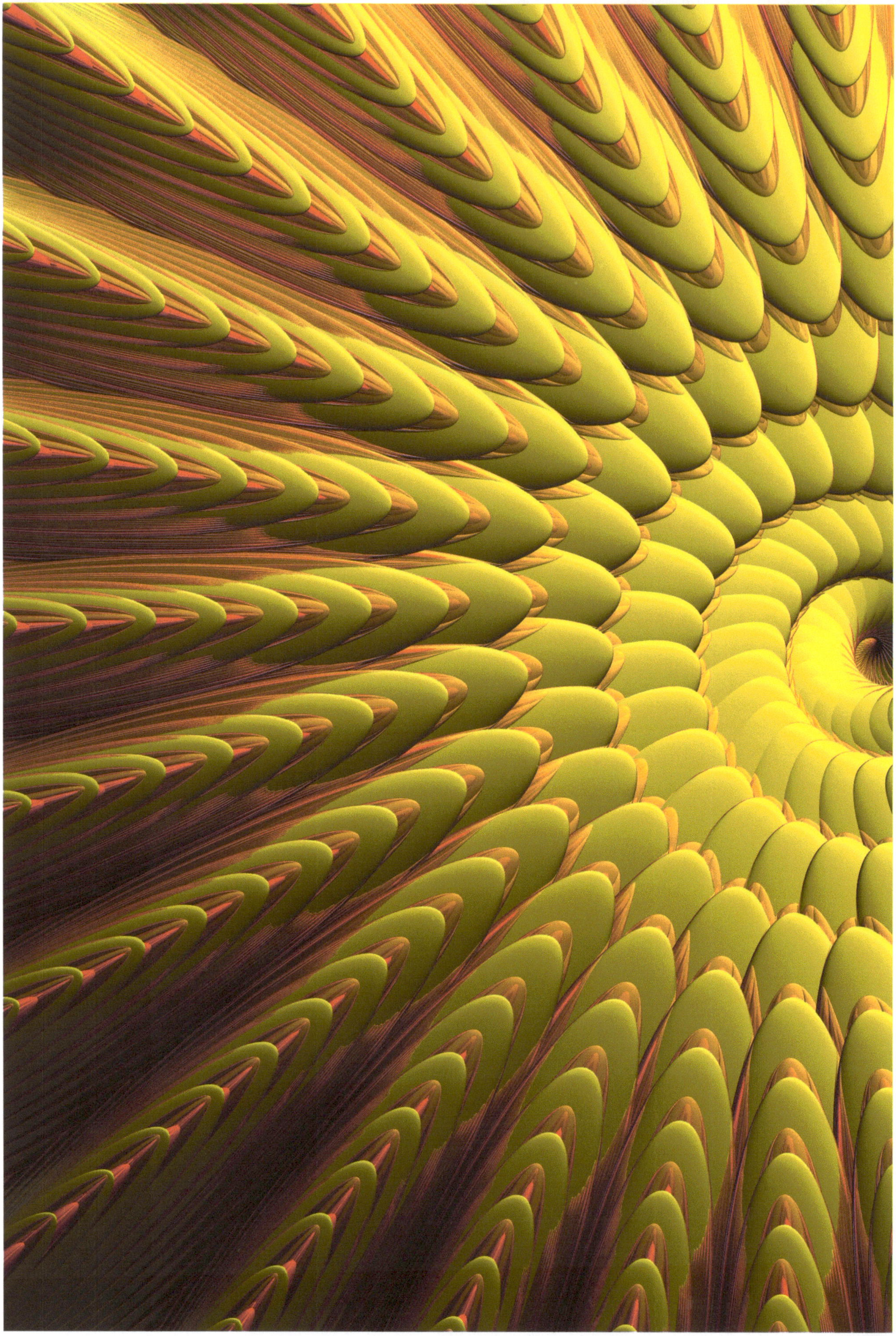

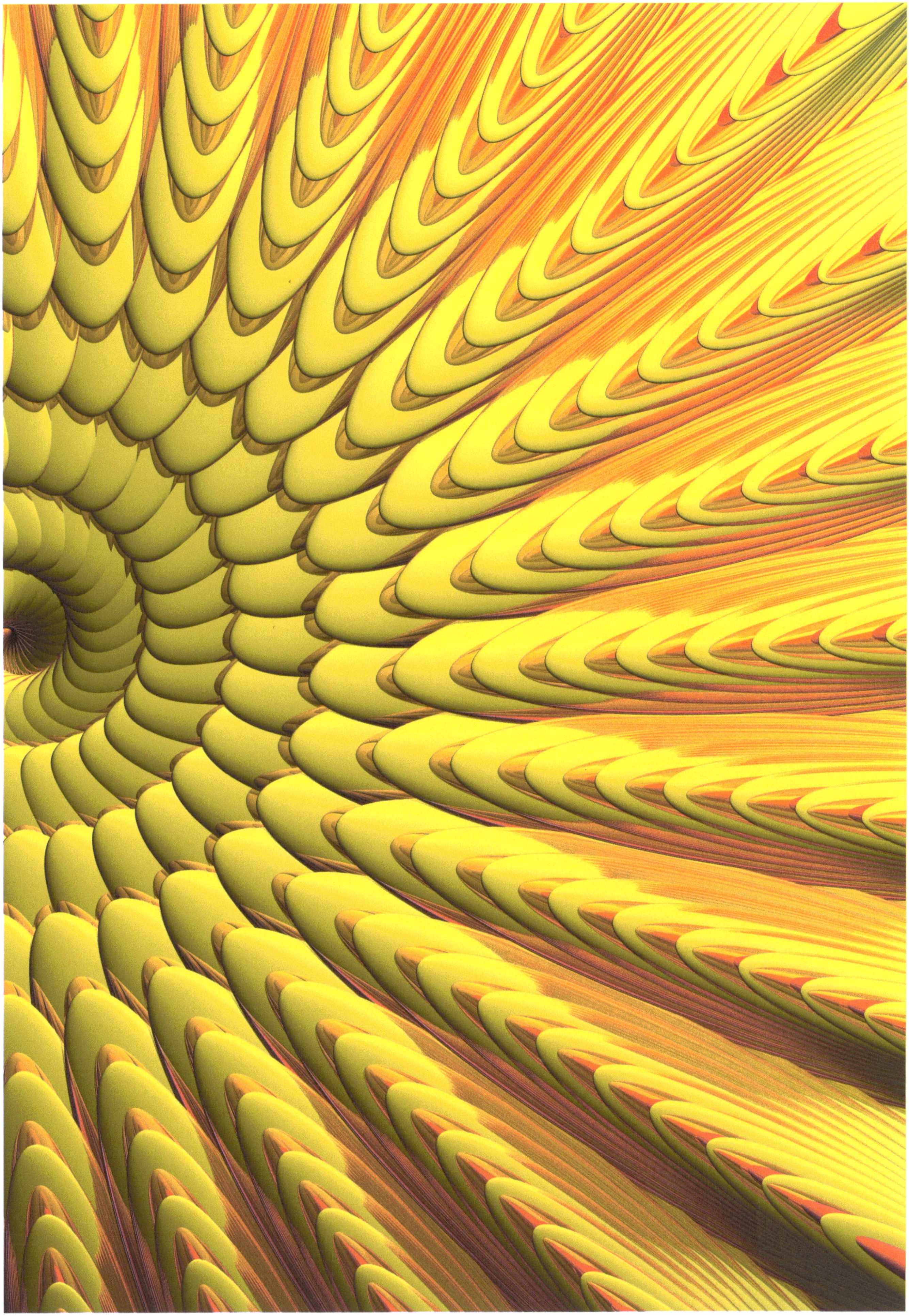

Previous pages:

STAR SPIRAL

DEAL OF HEARTS

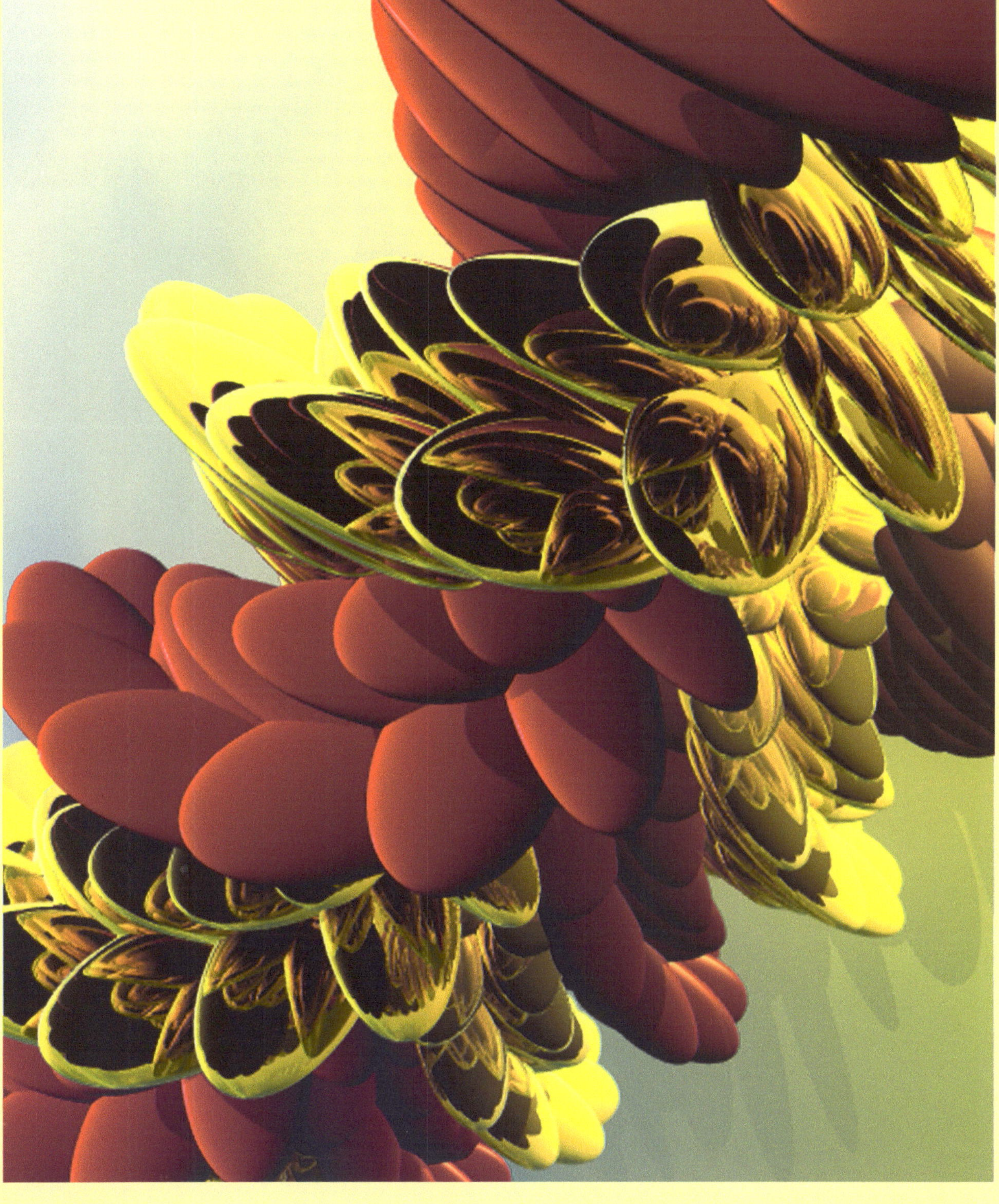

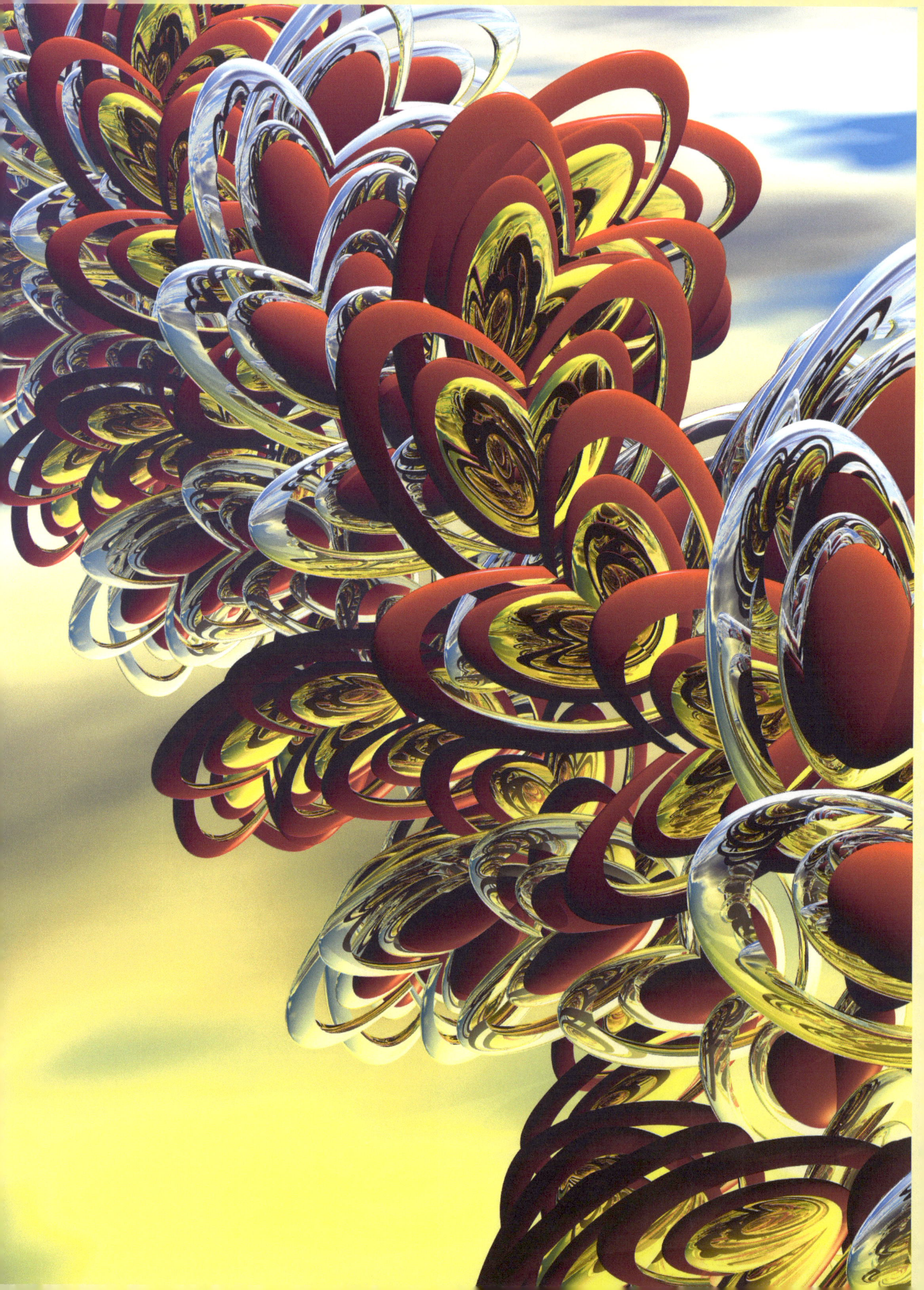

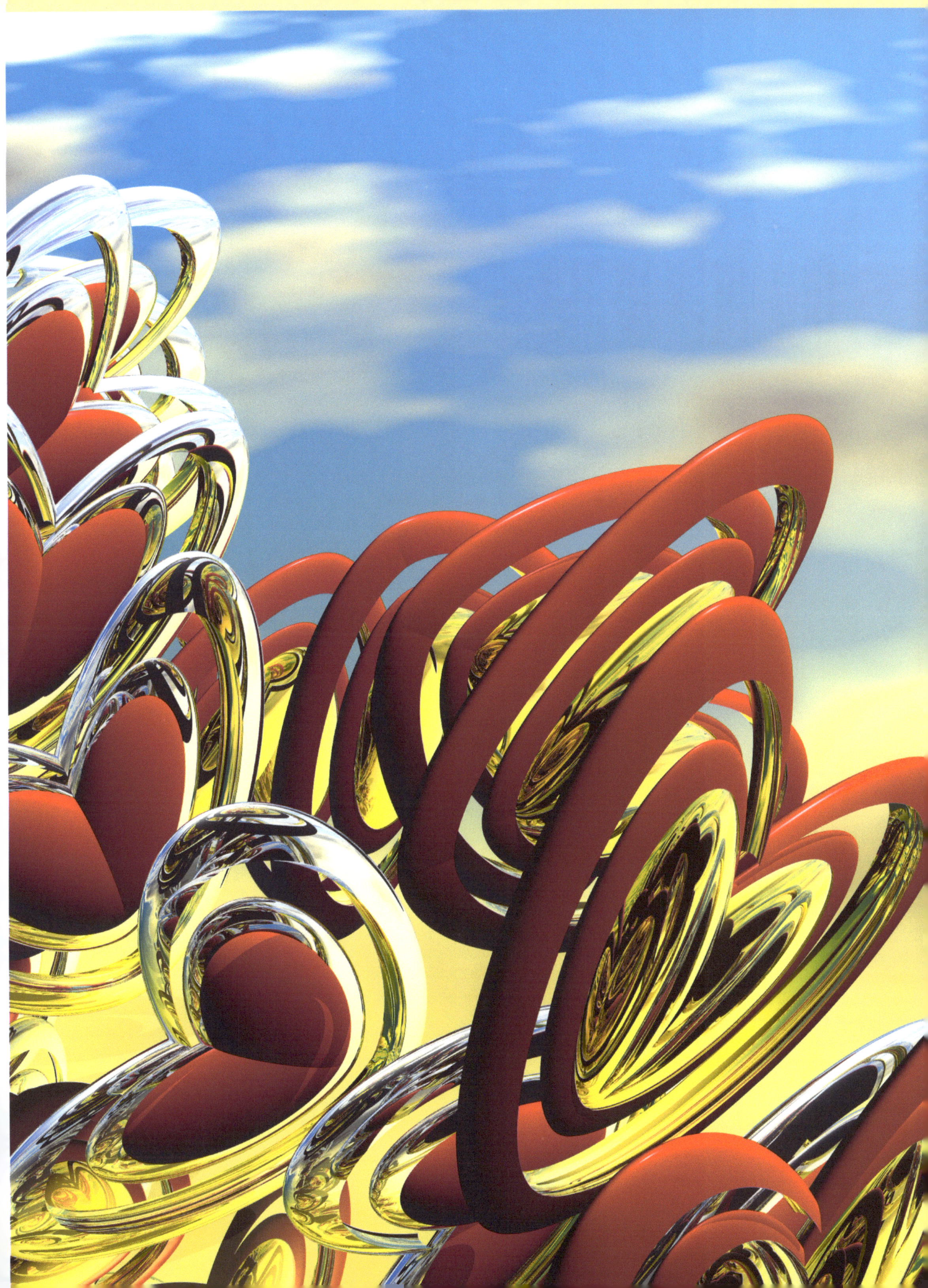

Previous pages:

COMPLEX HEART

GOLDEN TUBEZ

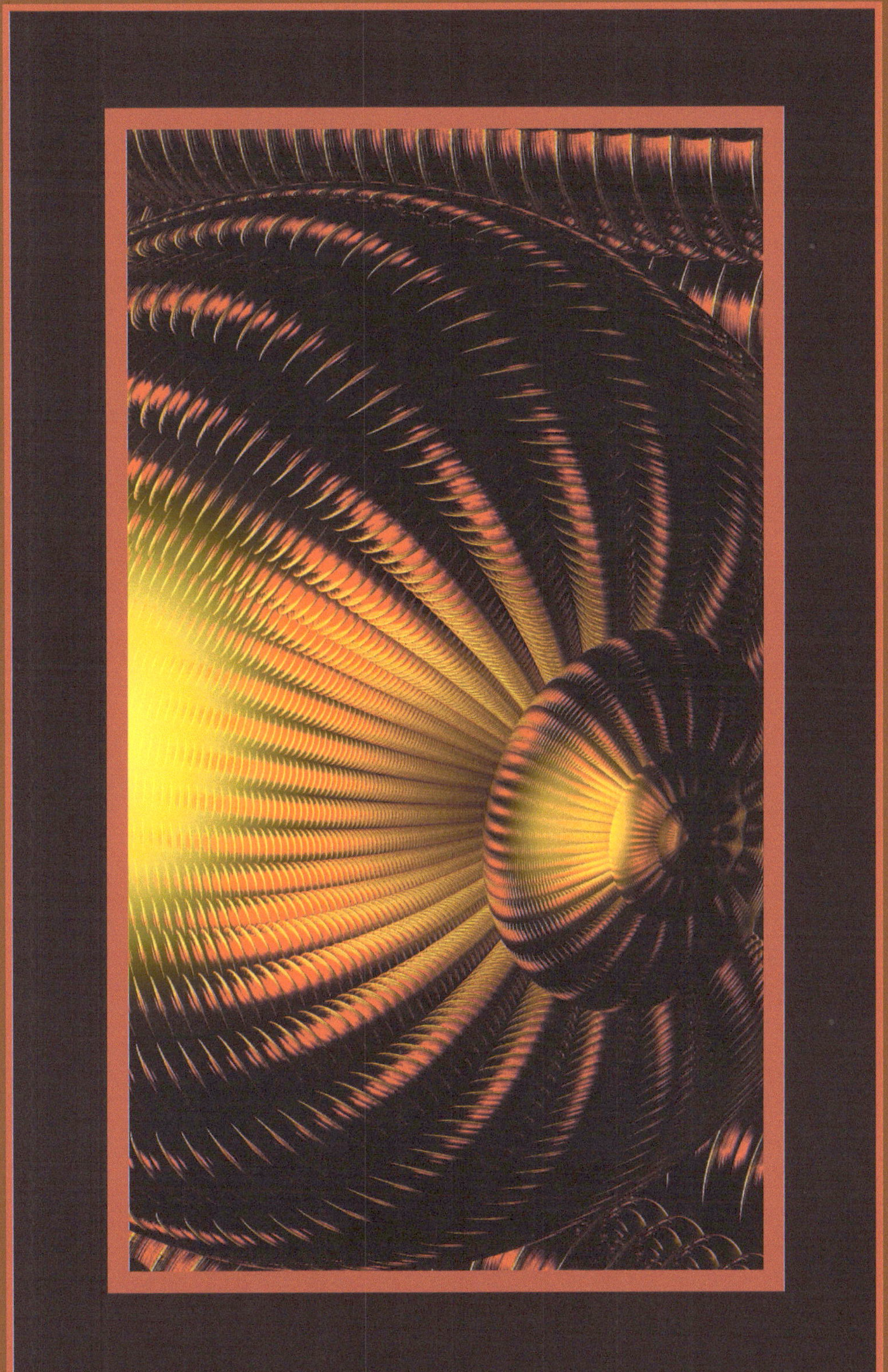

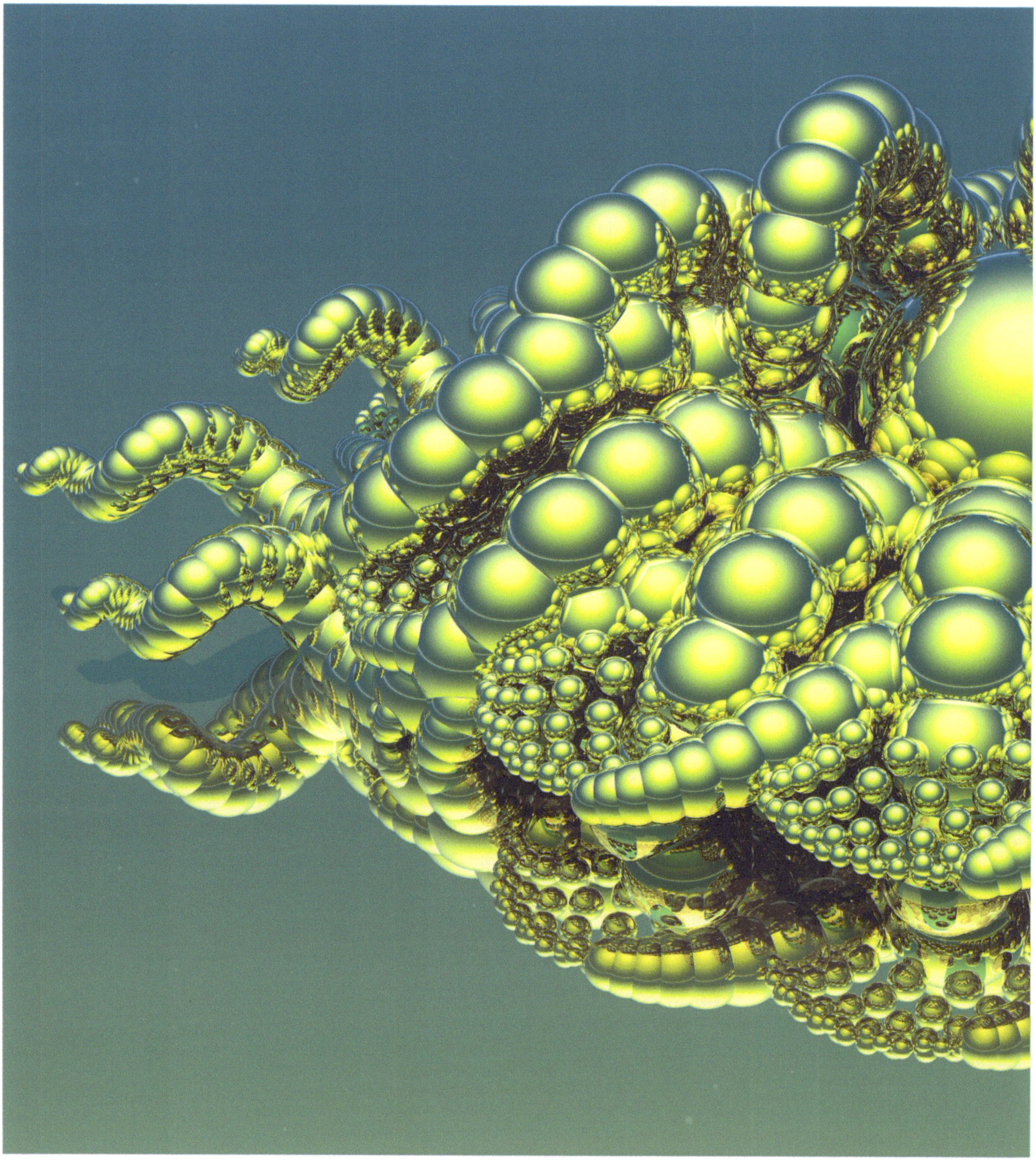

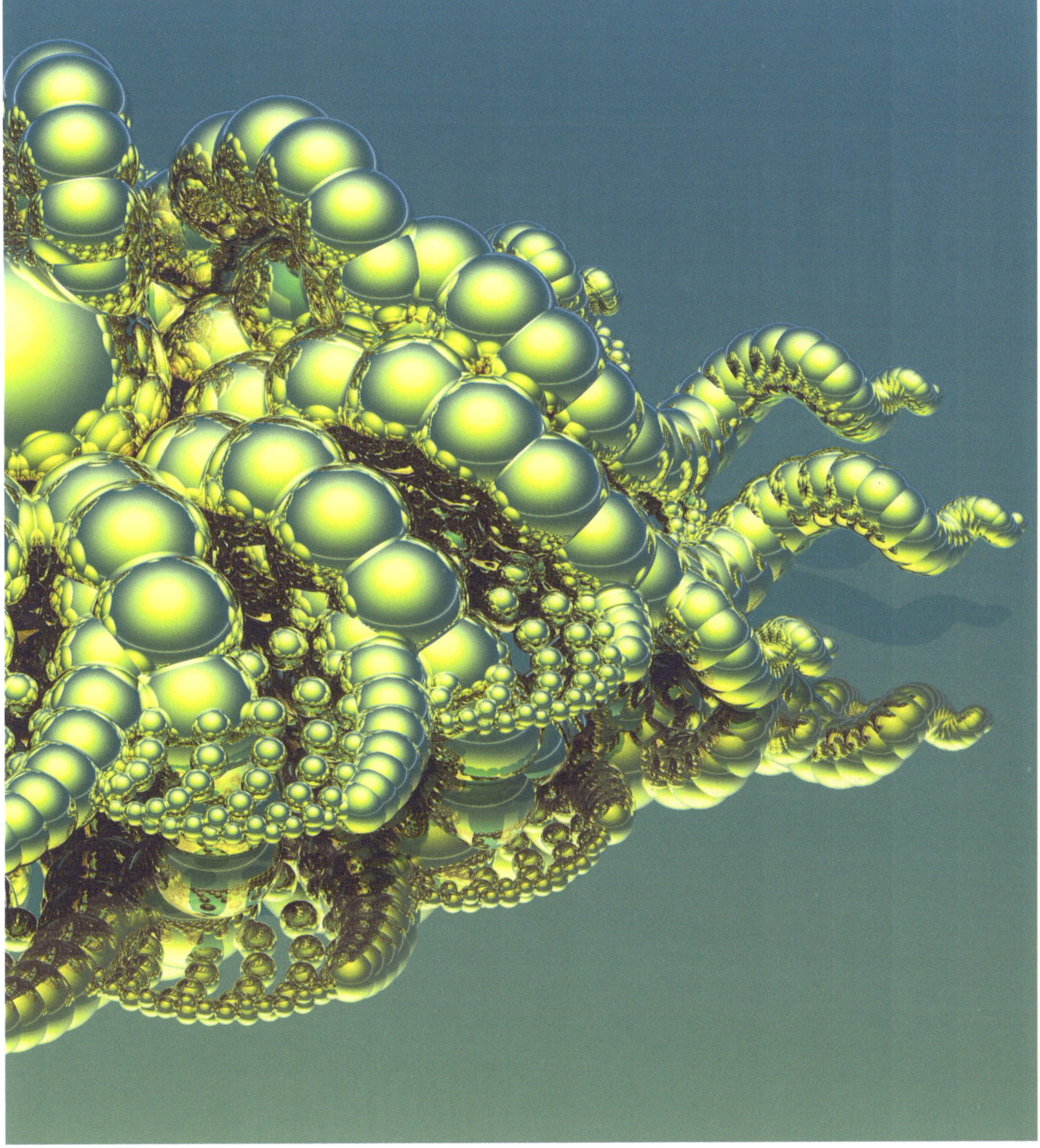

Previous pages:

LYDIA AND THE OCTOPUS

ORBSPHERE

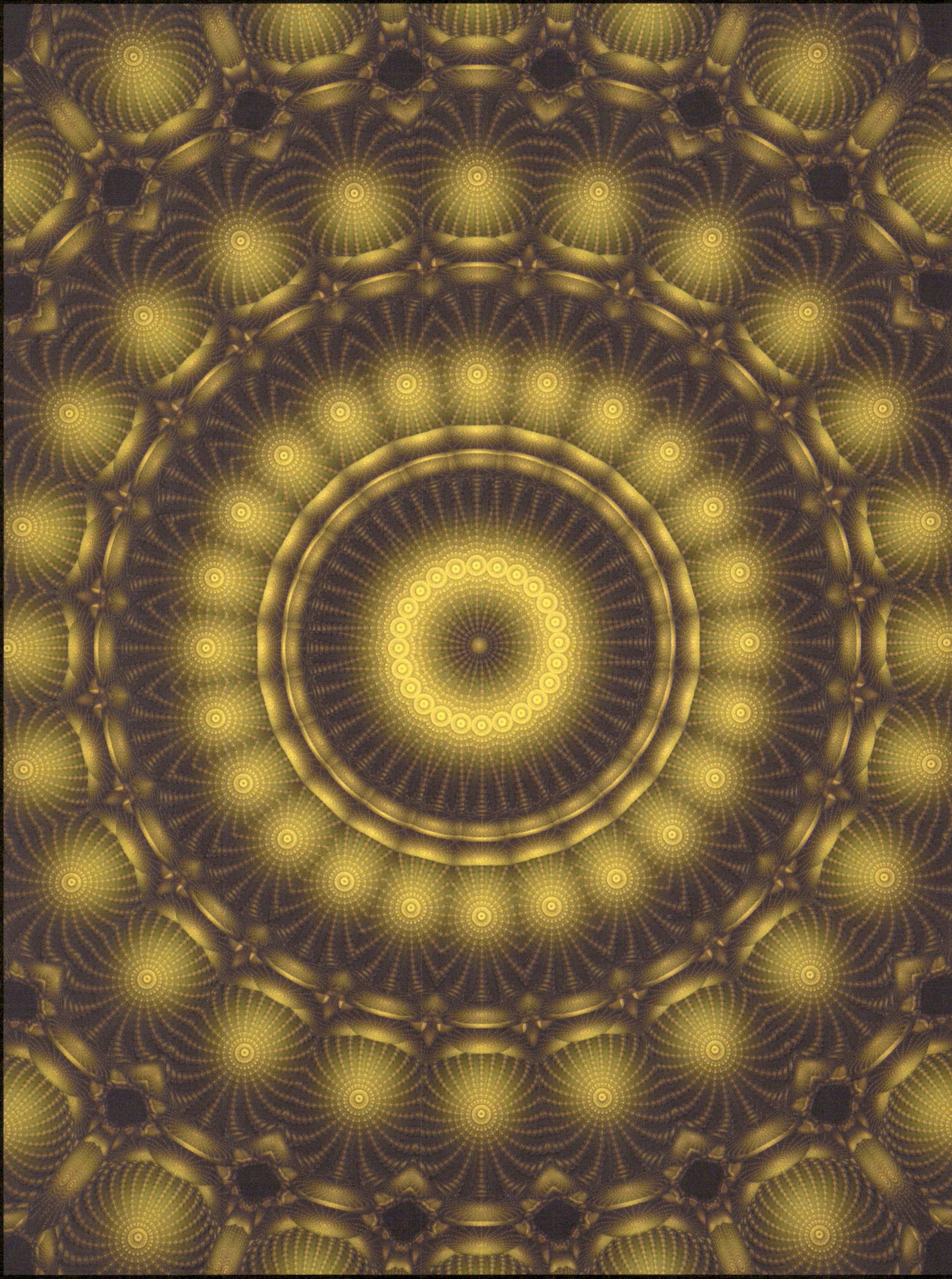

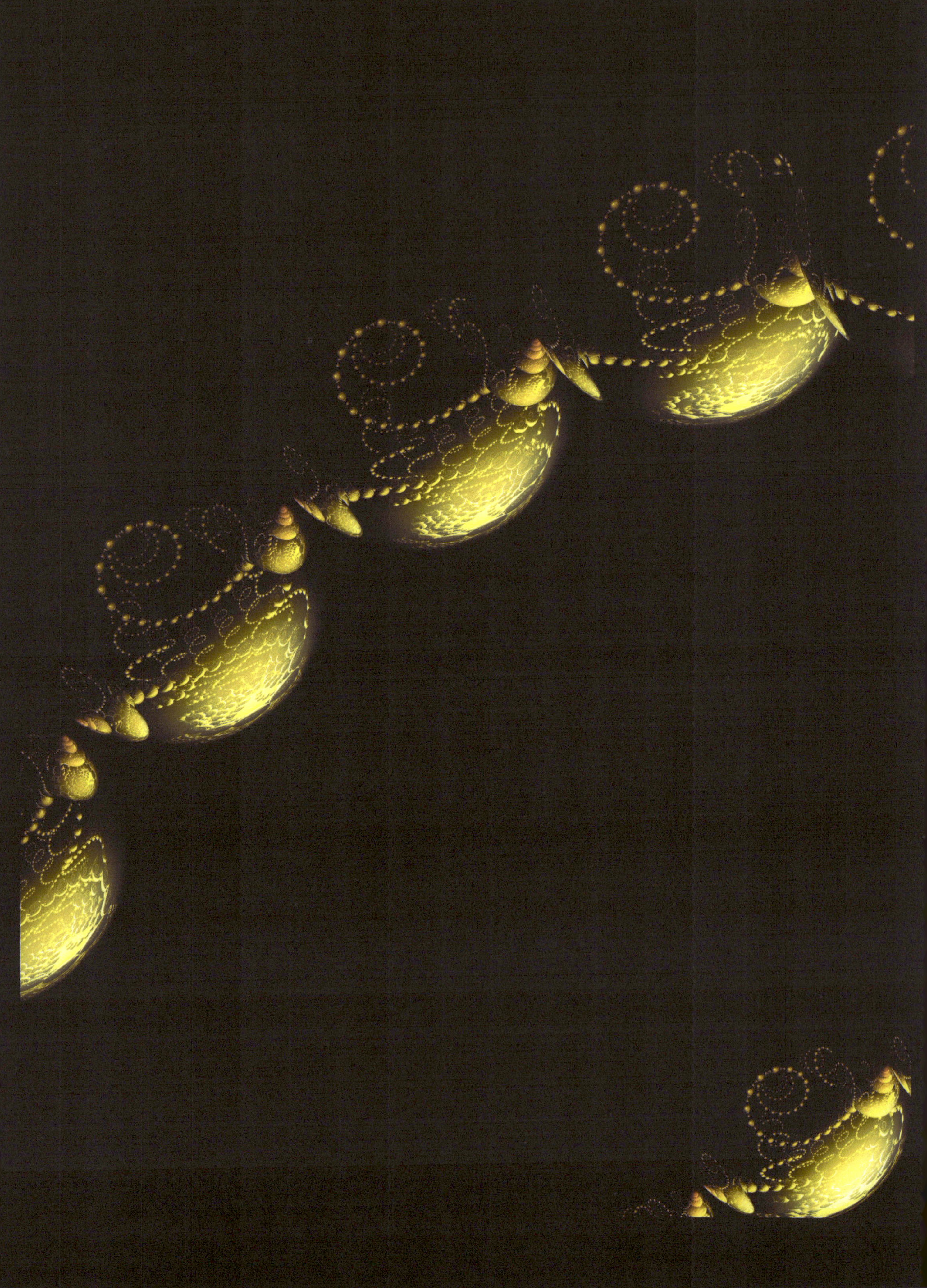

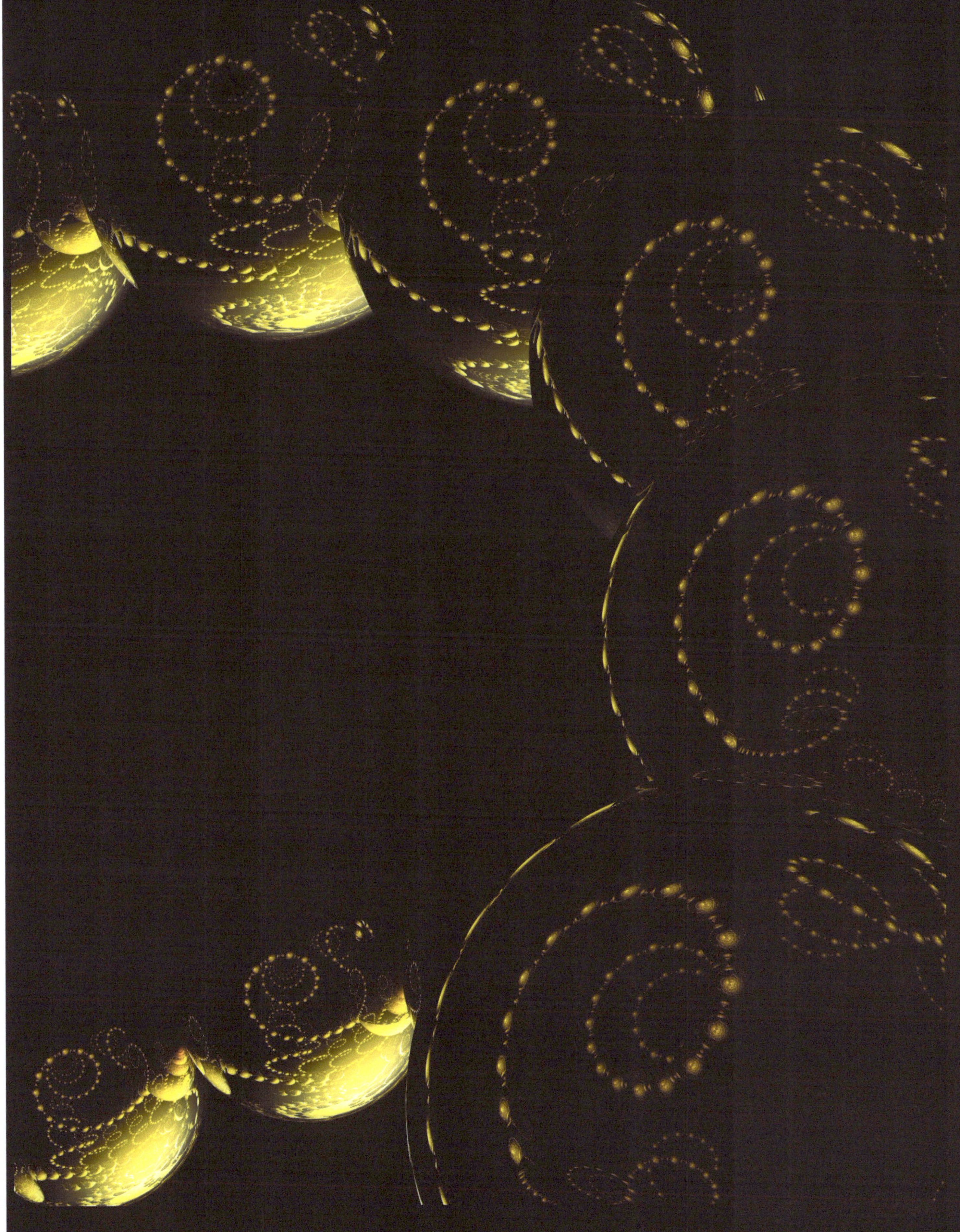

Previous pages:

100PUNIT

100PUNIT2

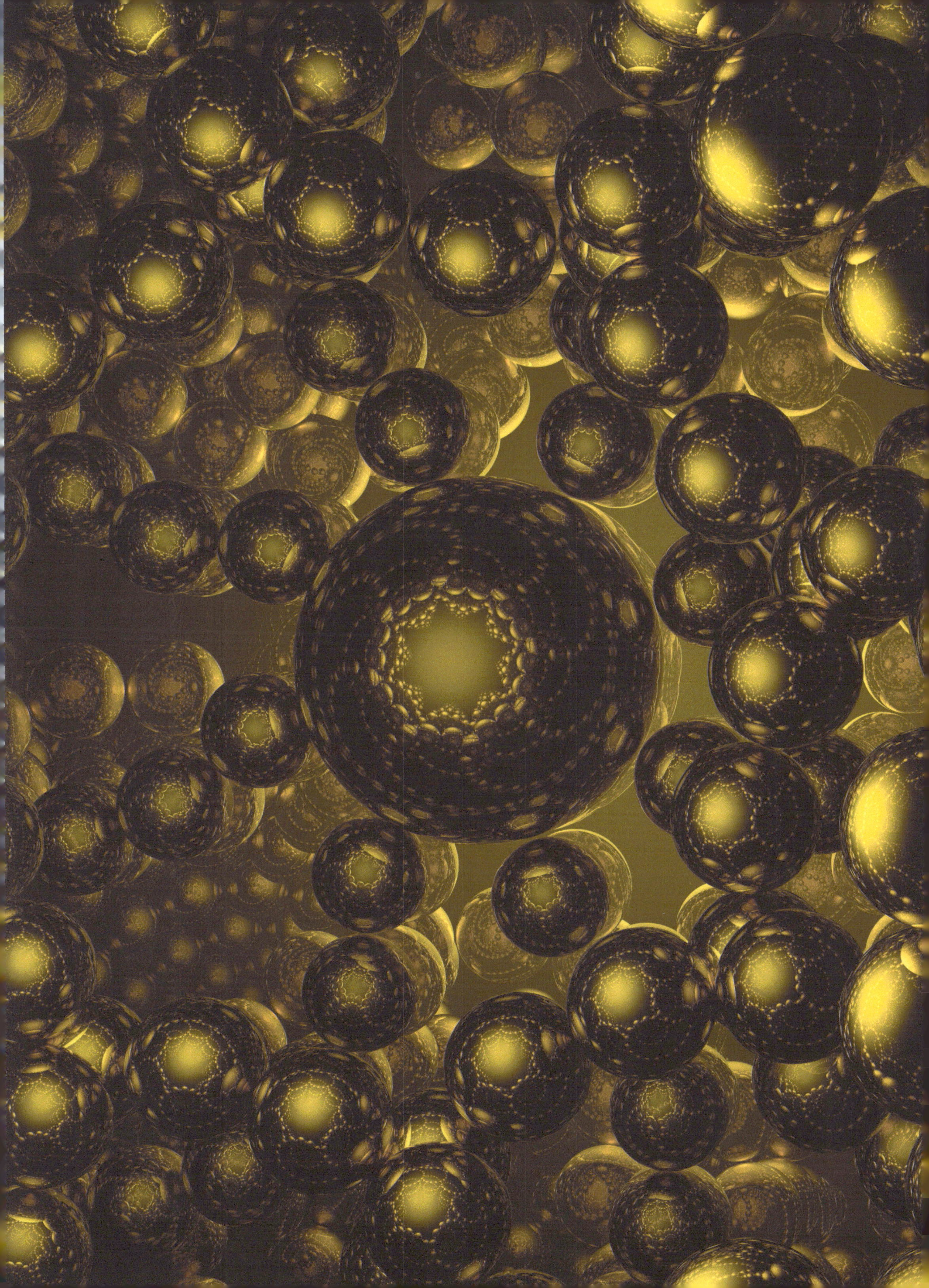

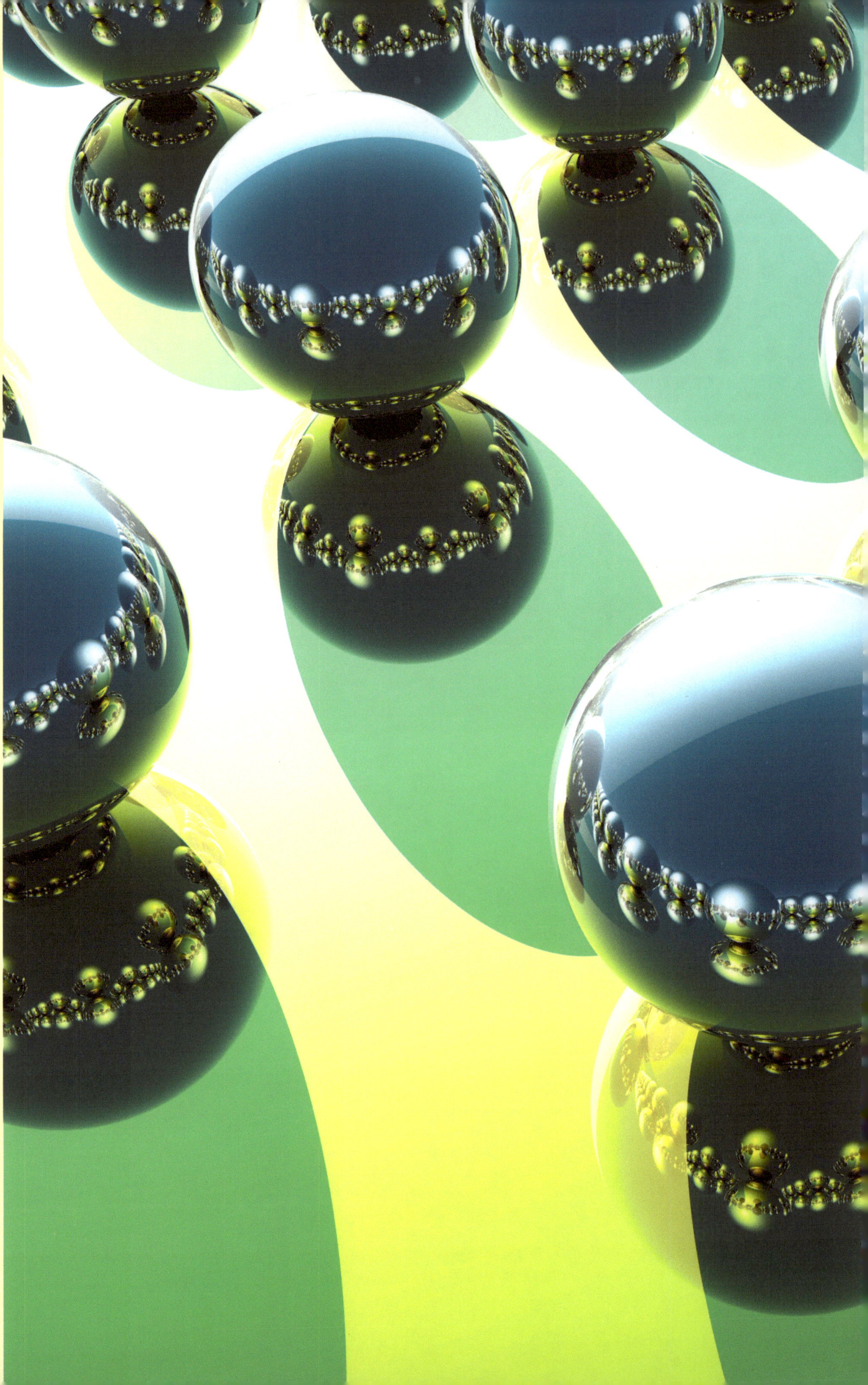

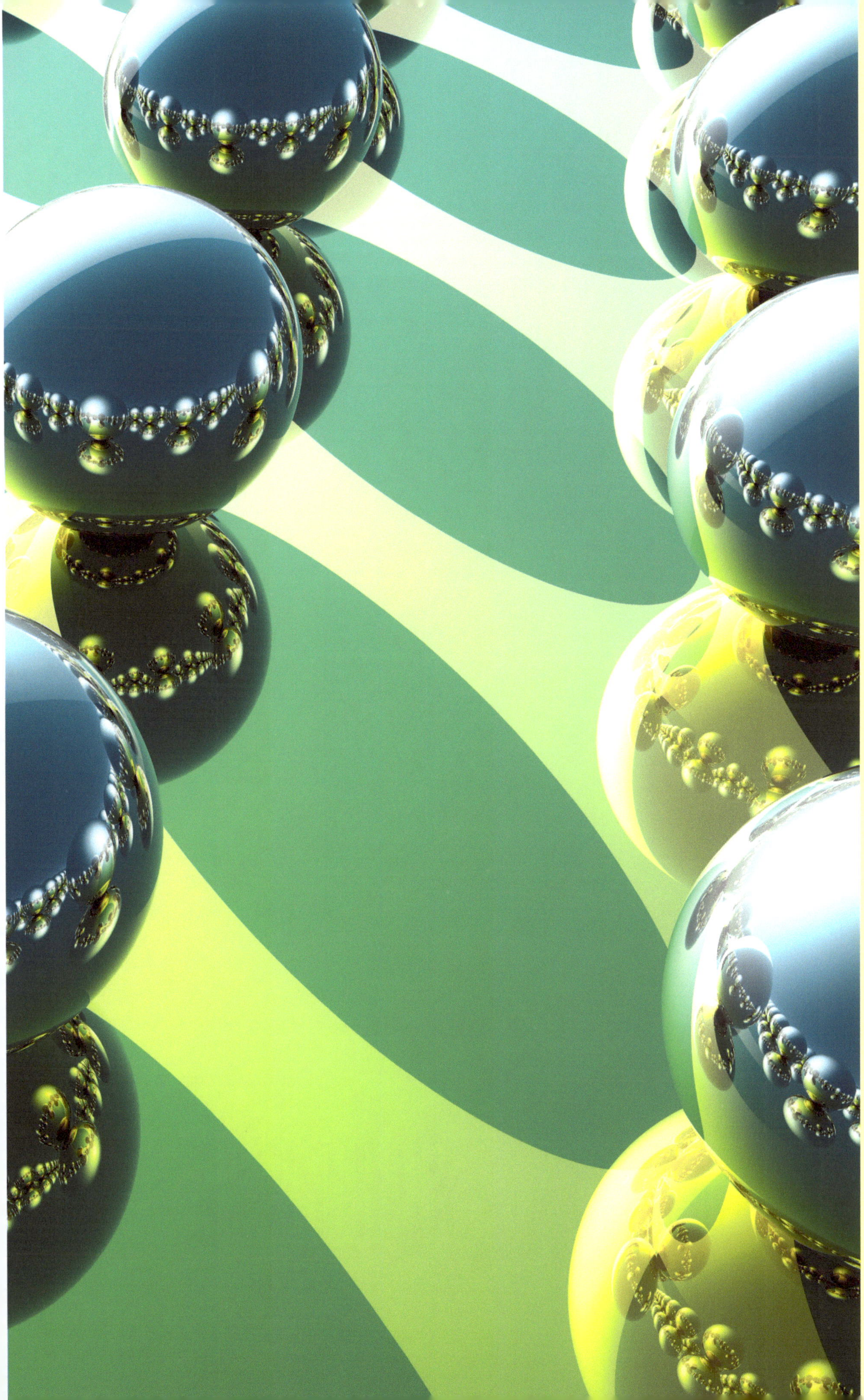

Previous pages:

SPHEREFIELD

STARCHILDREN

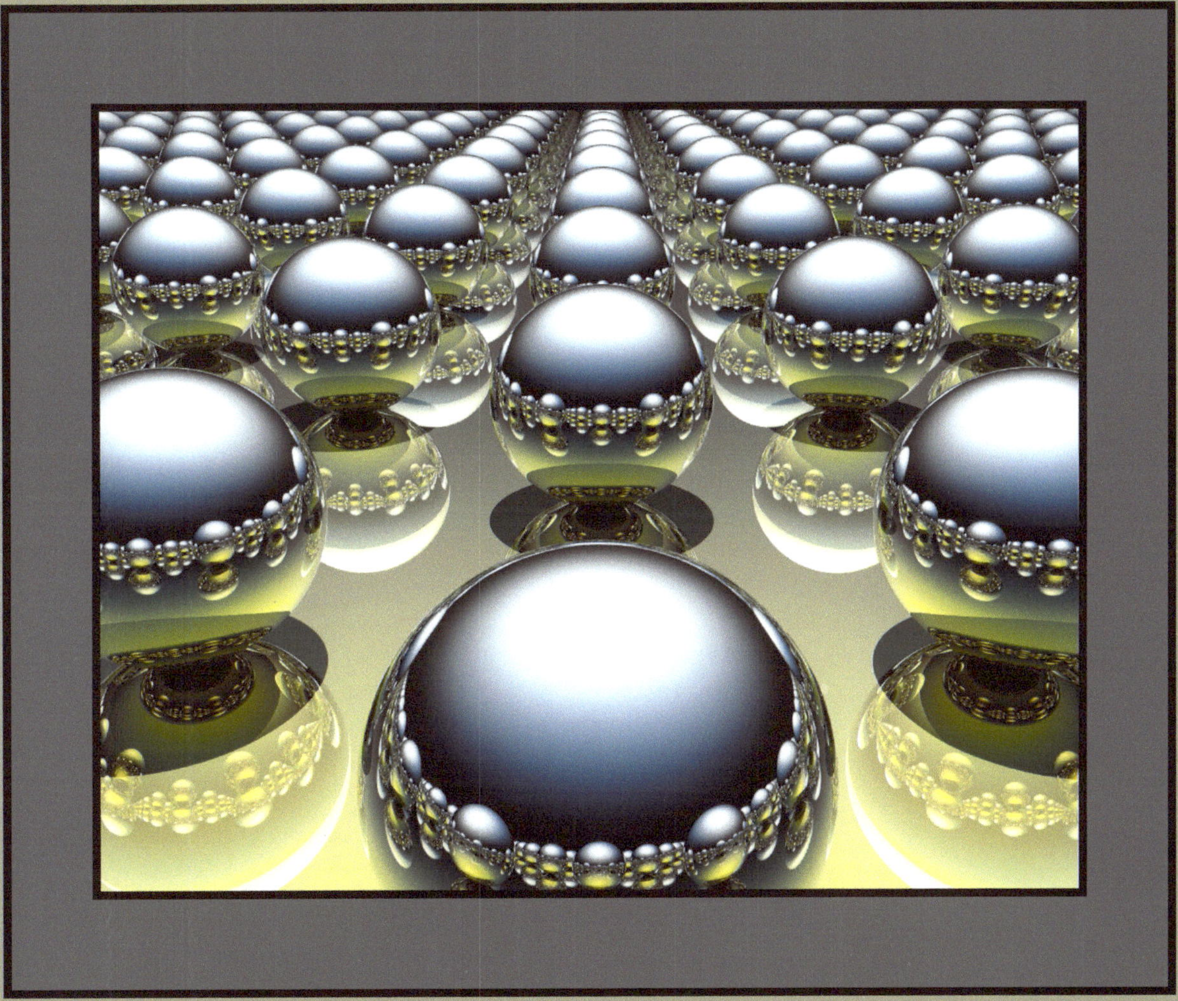

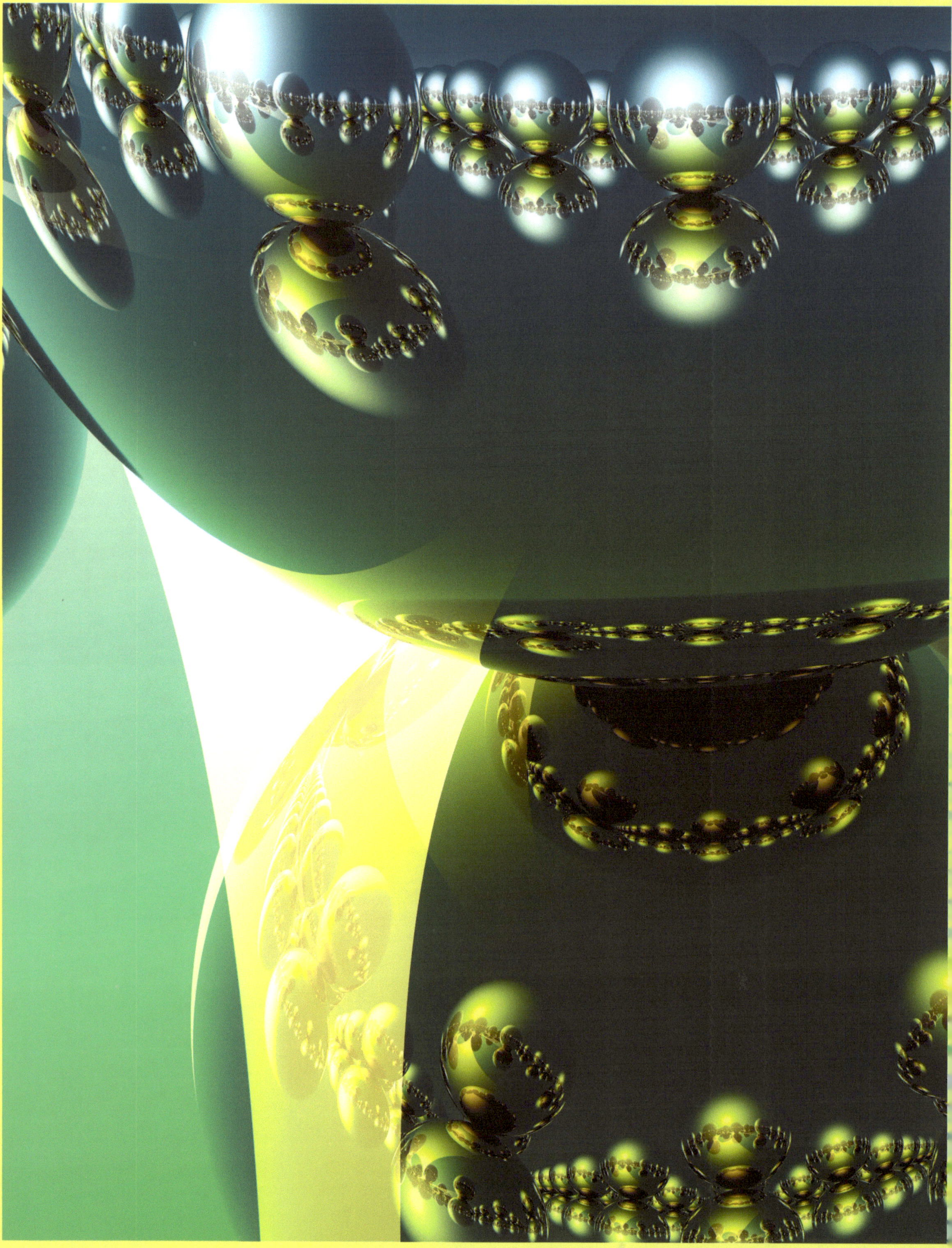

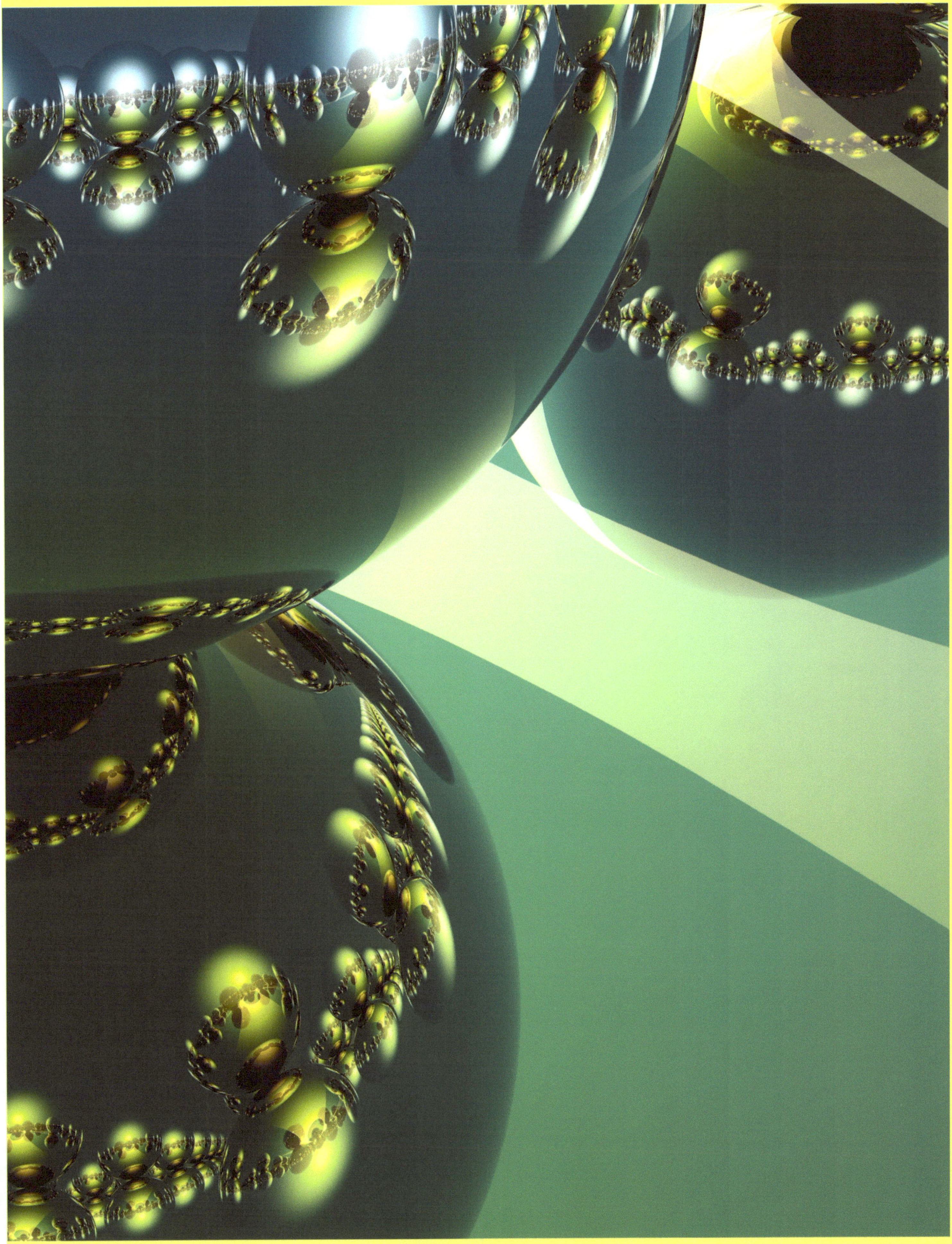

Previous pages:

STAR CHILD

A GOOD THOUGHT

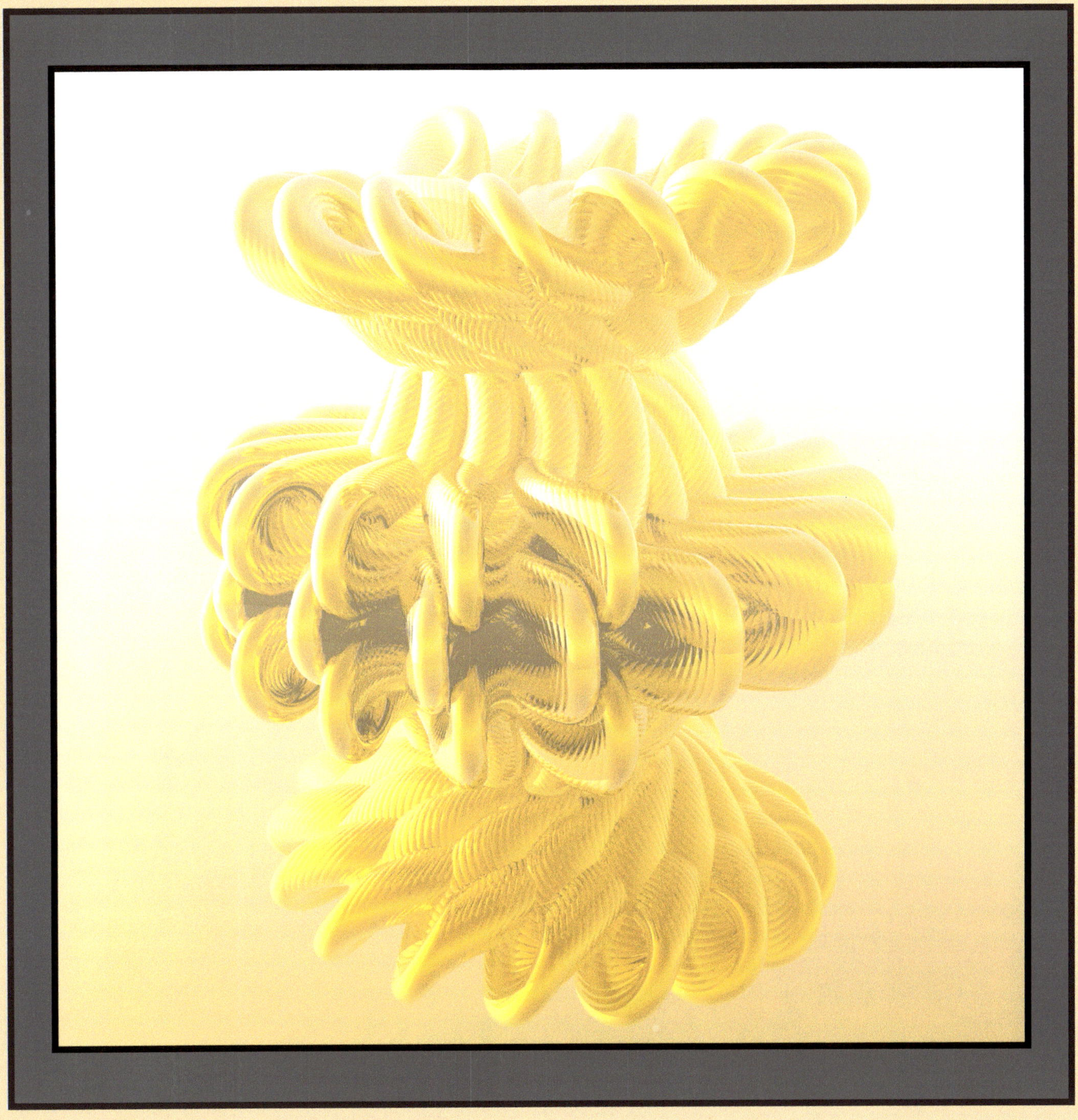

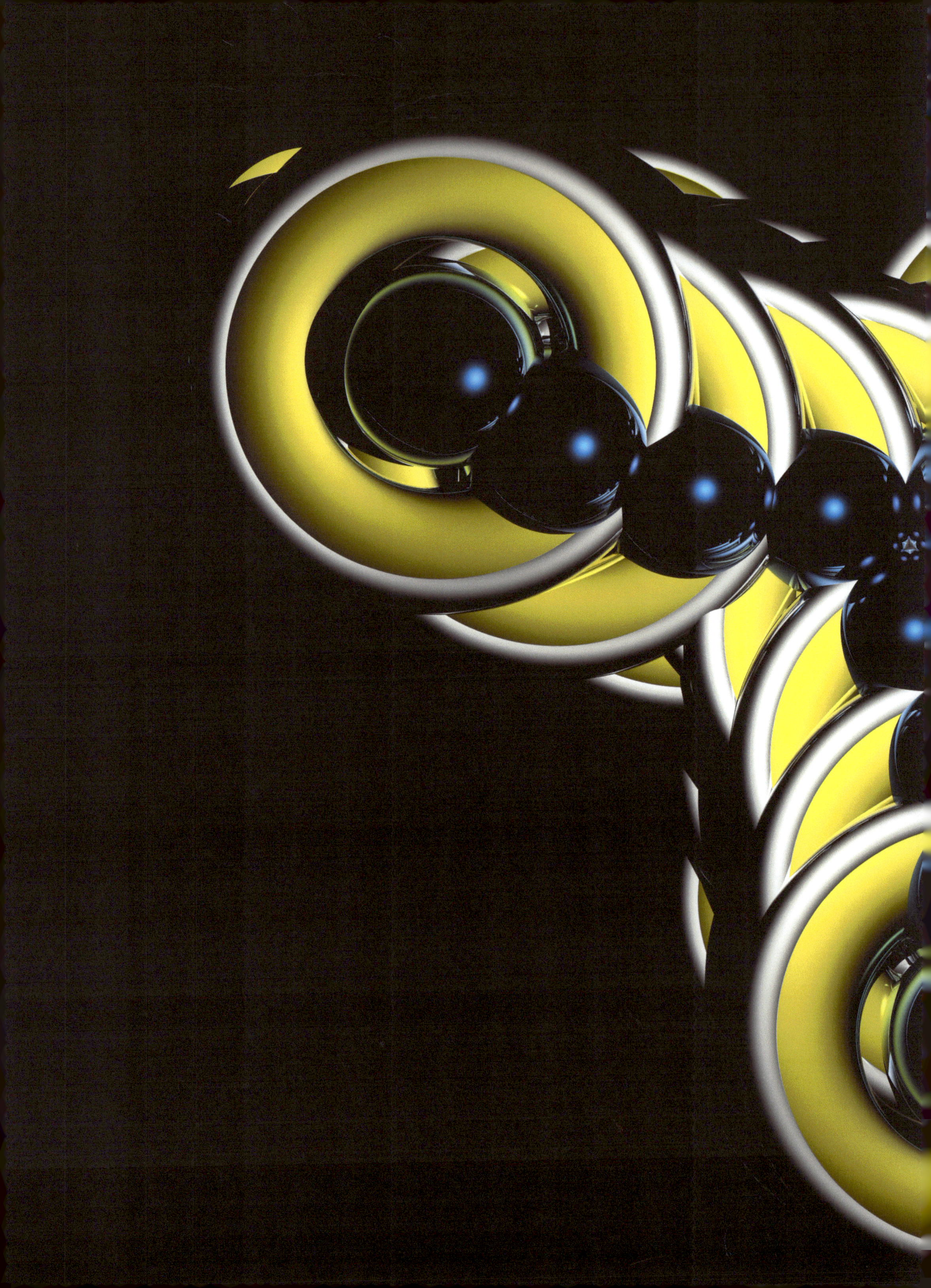

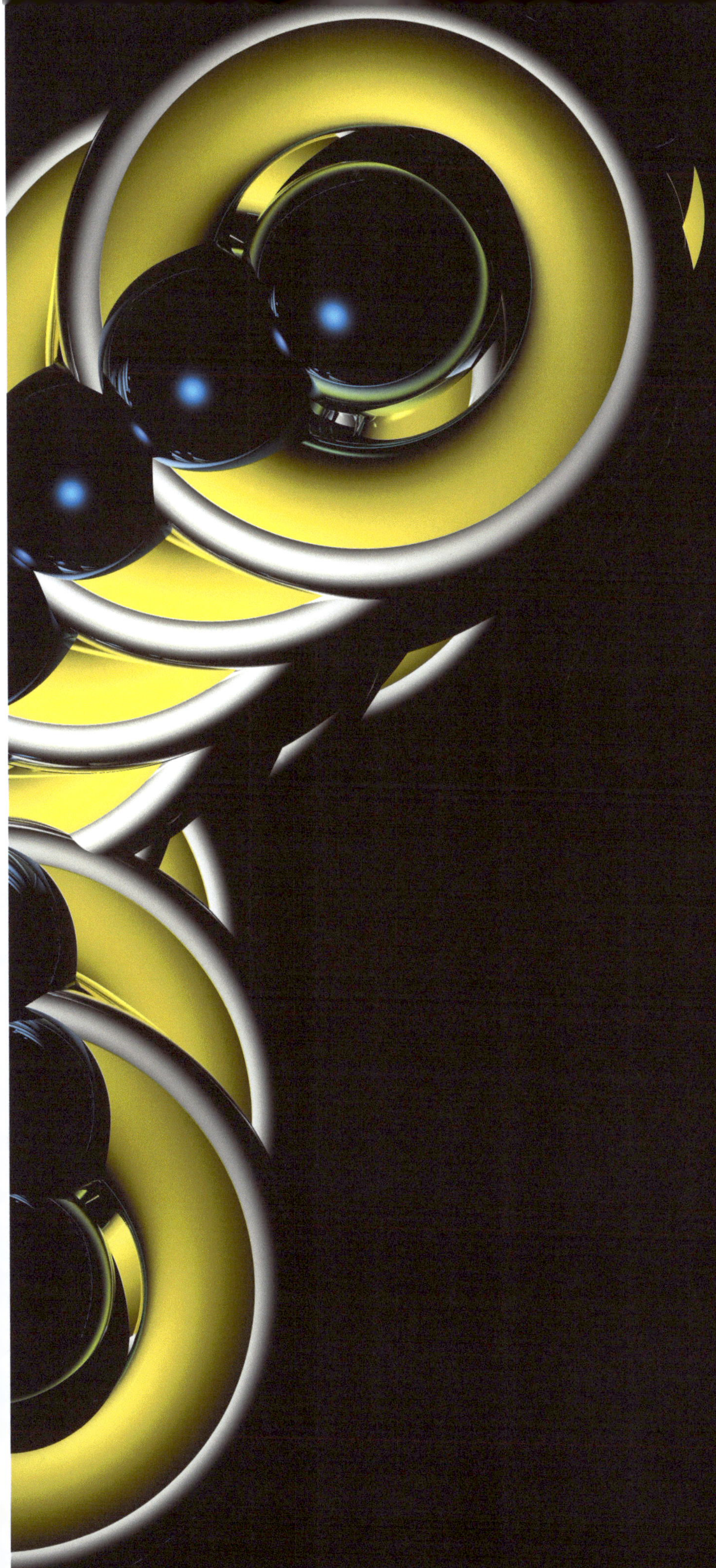

Previous pages:

DARK COILER

BEAUTY

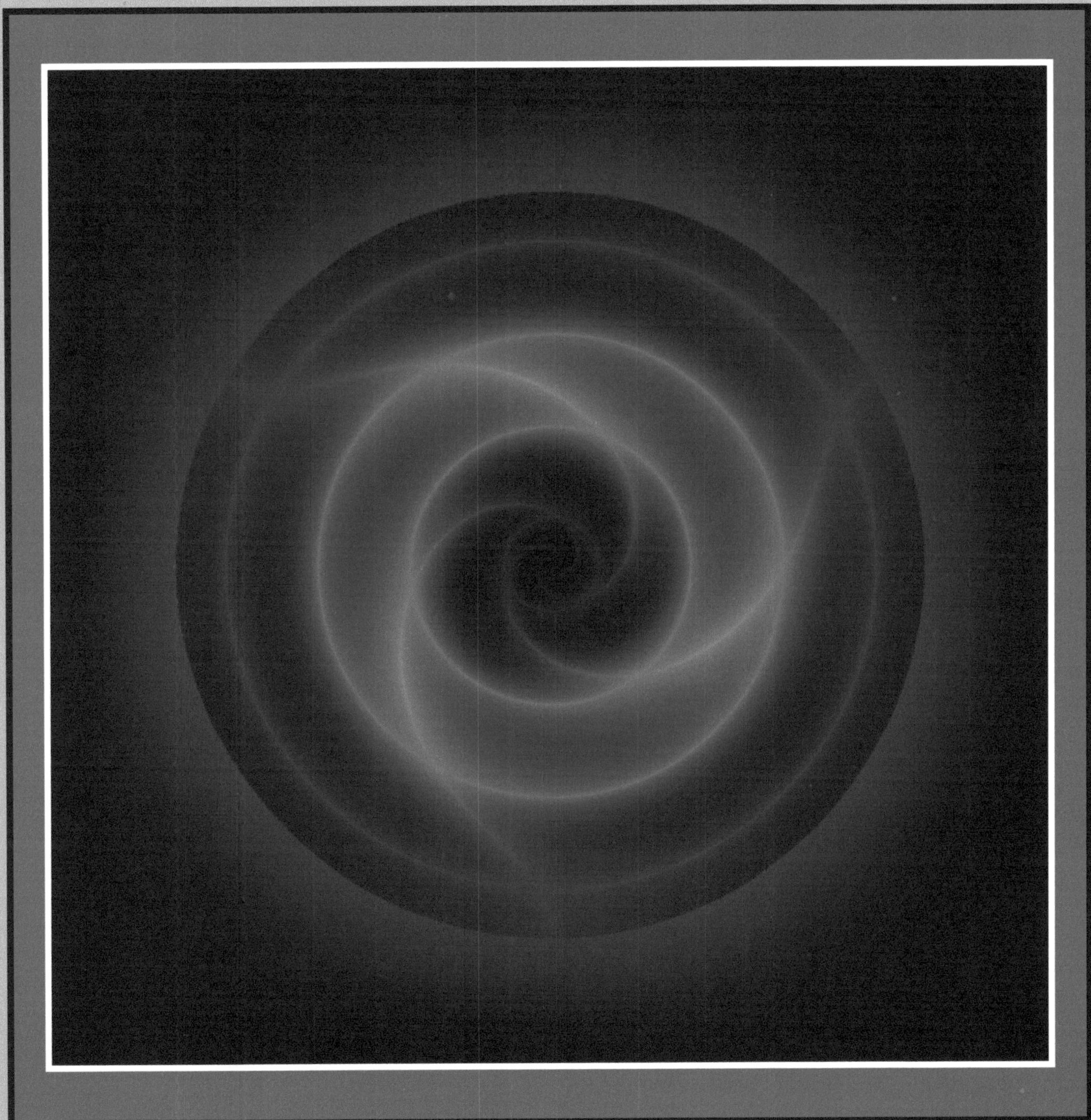

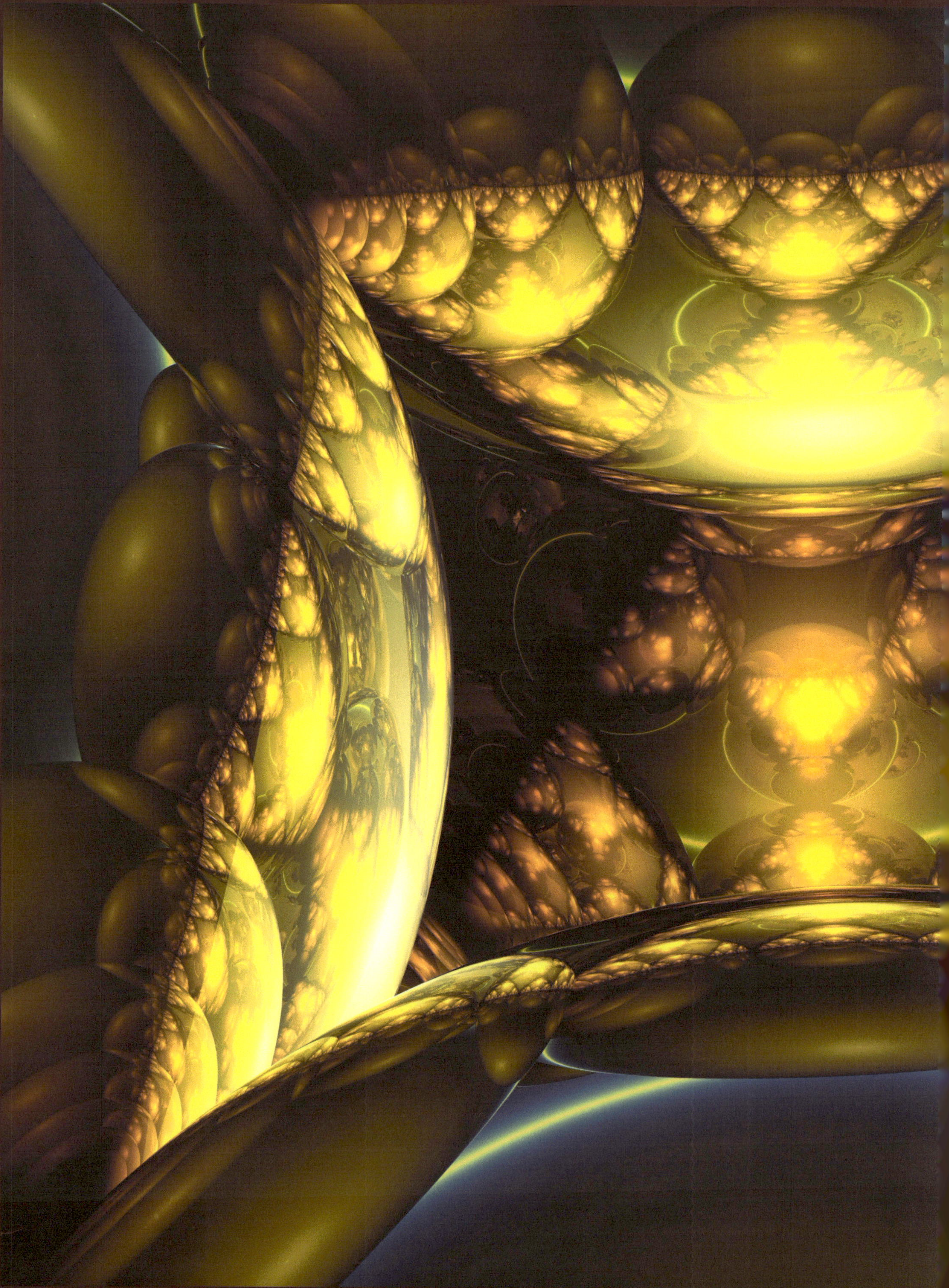

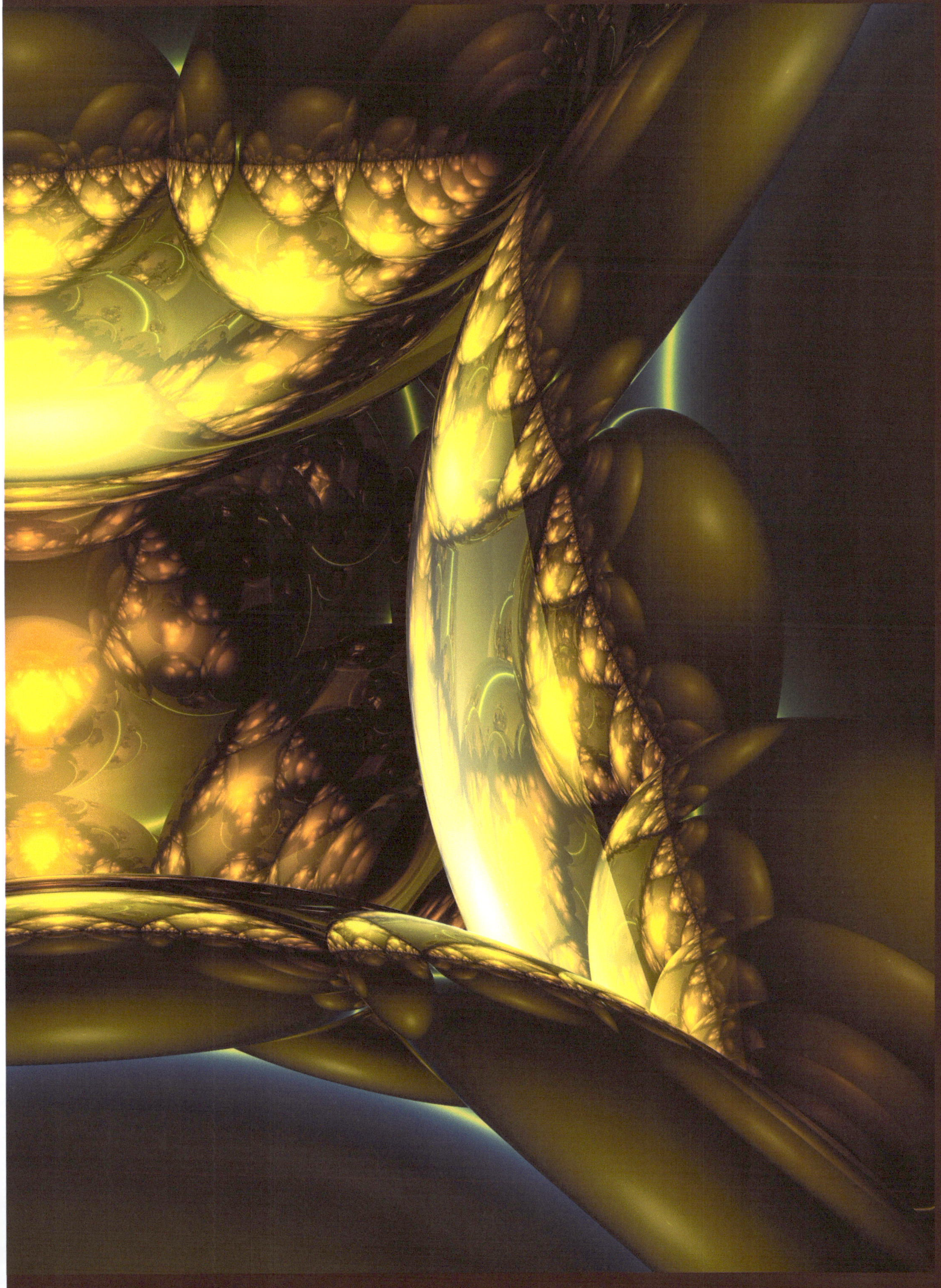

Previous pages:

YARC

FRACTACIOUS

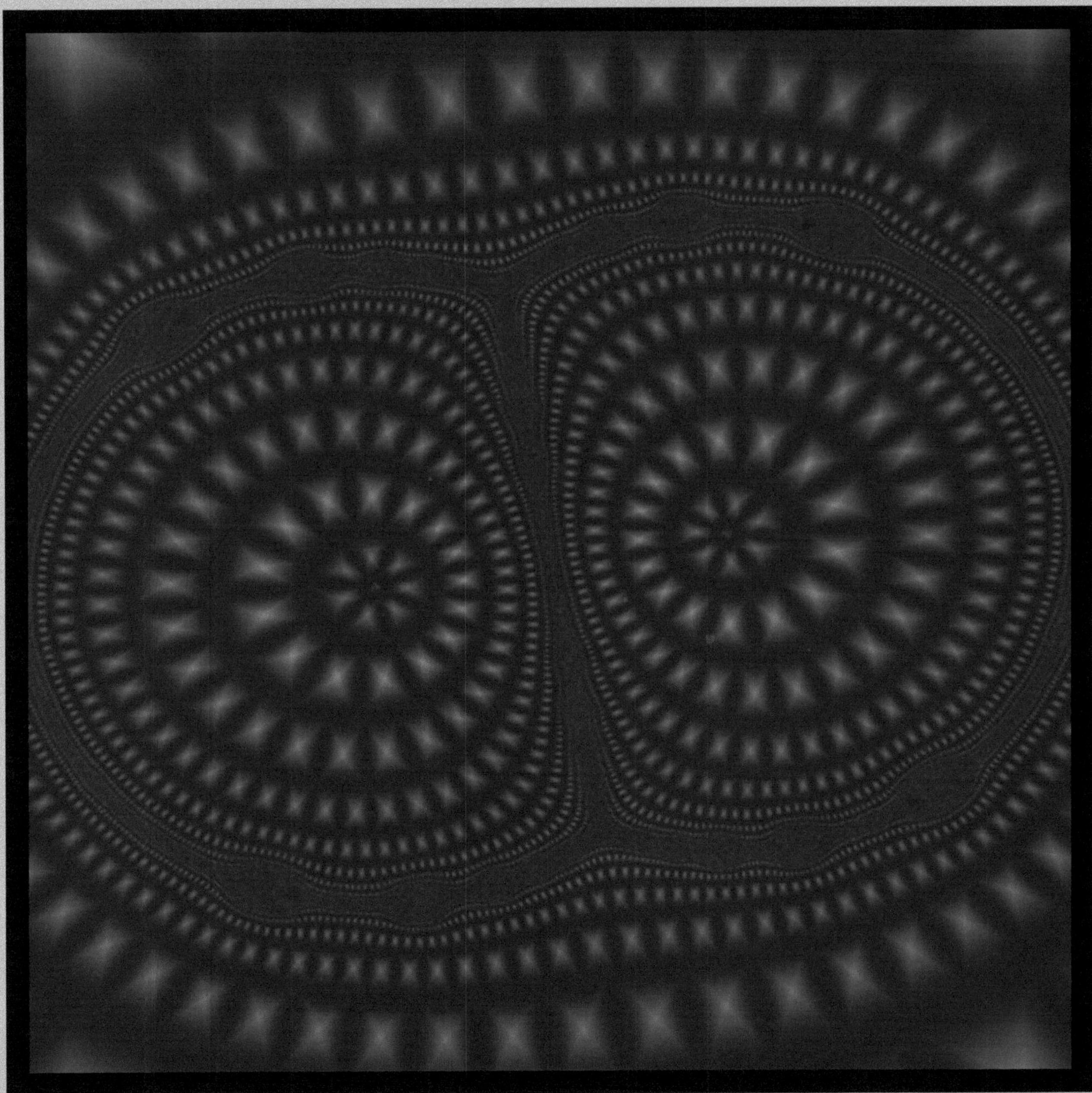

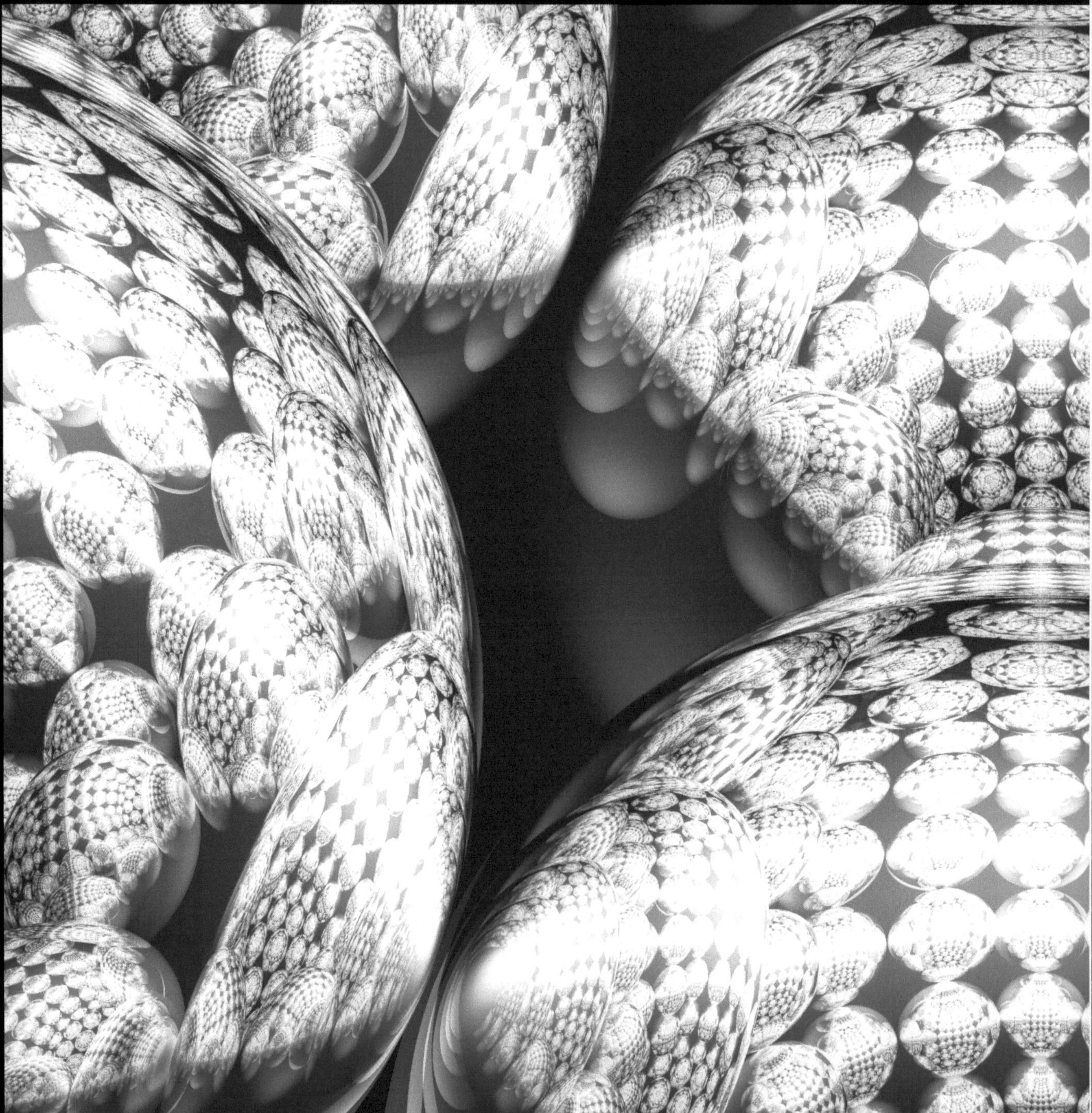

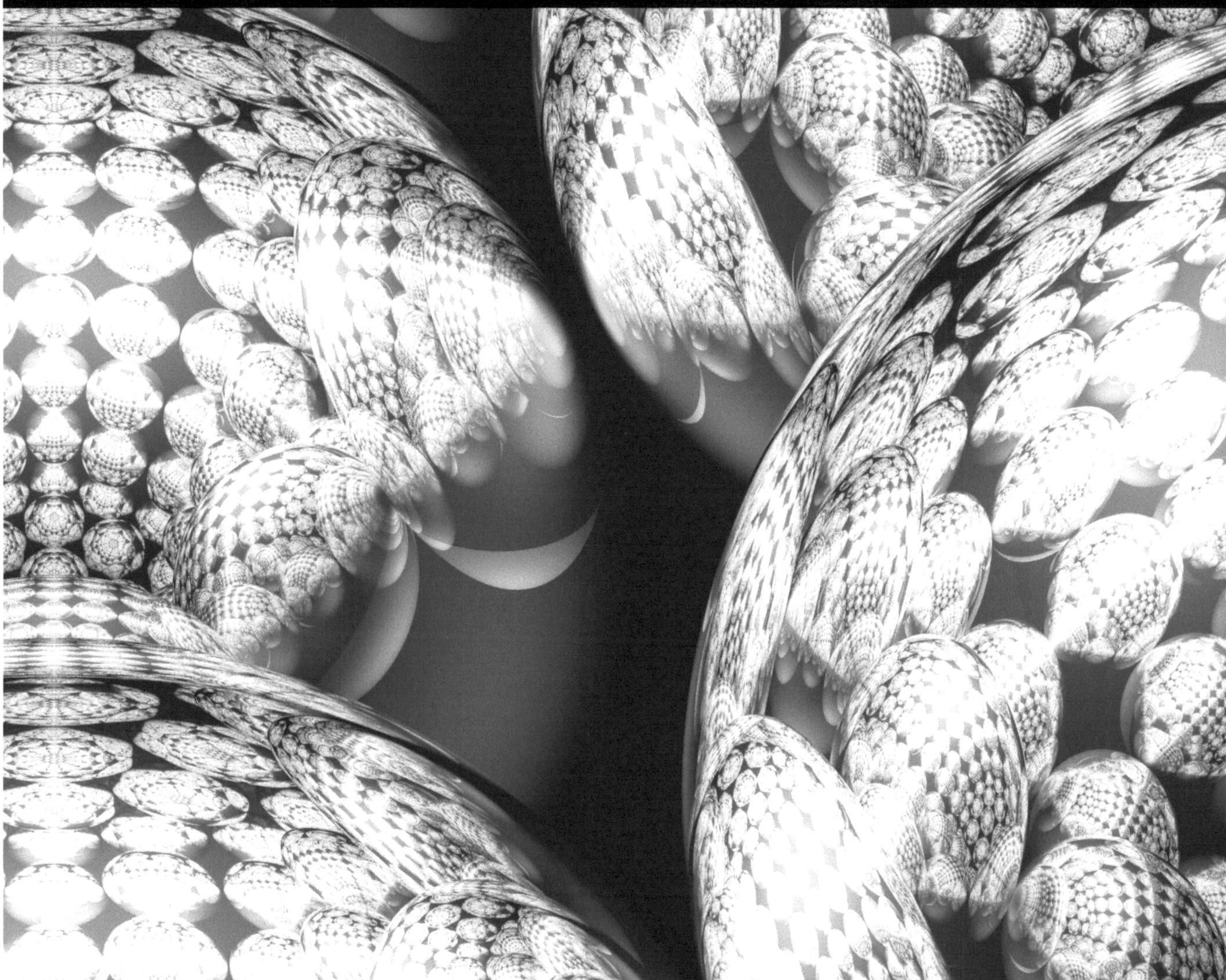

Previous pages:

NANOTUBE

WORLZ SILVER

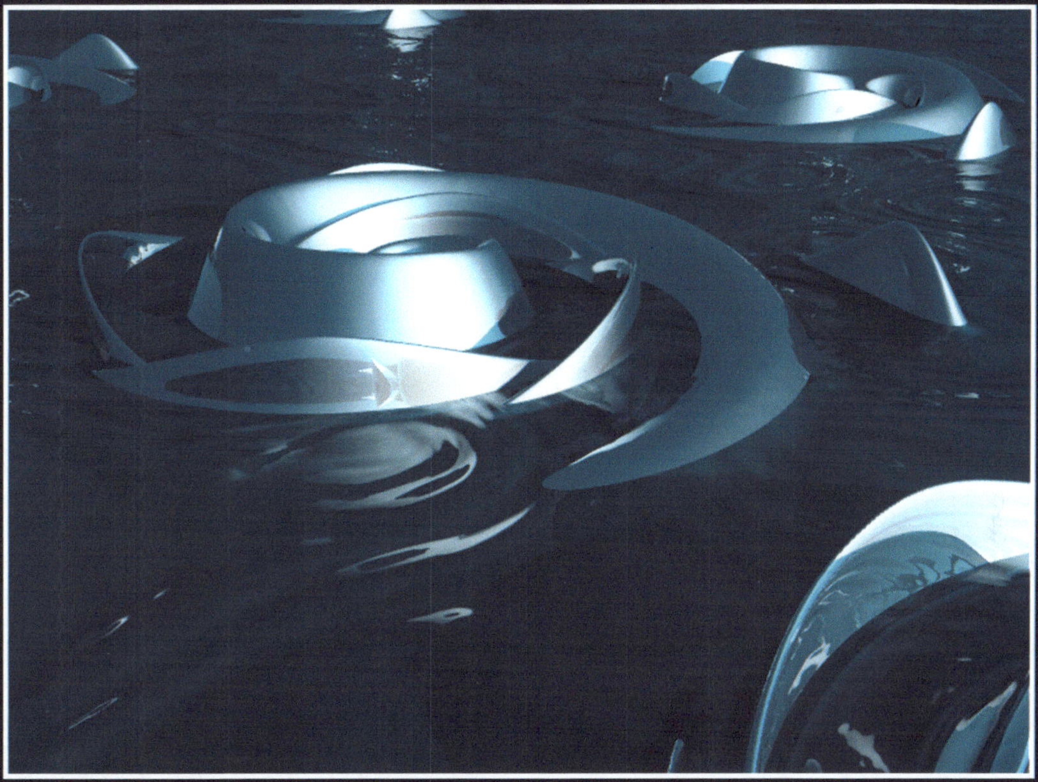

QUARTET

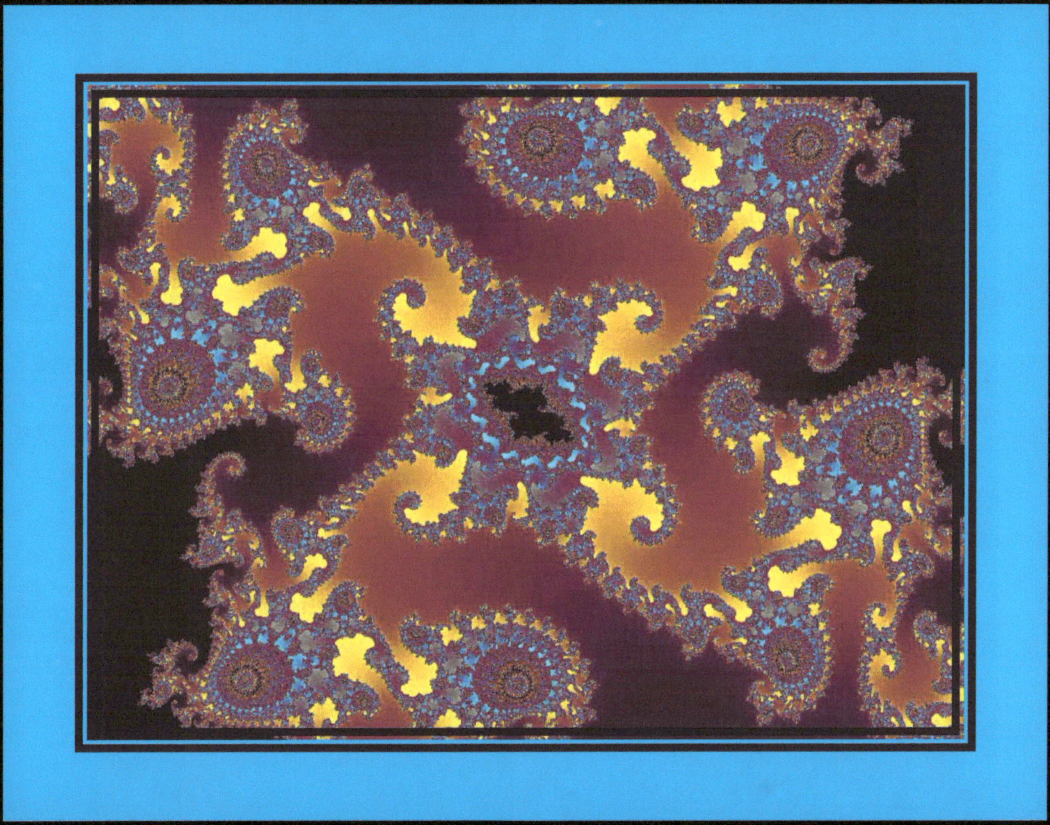

WARPED METALIX

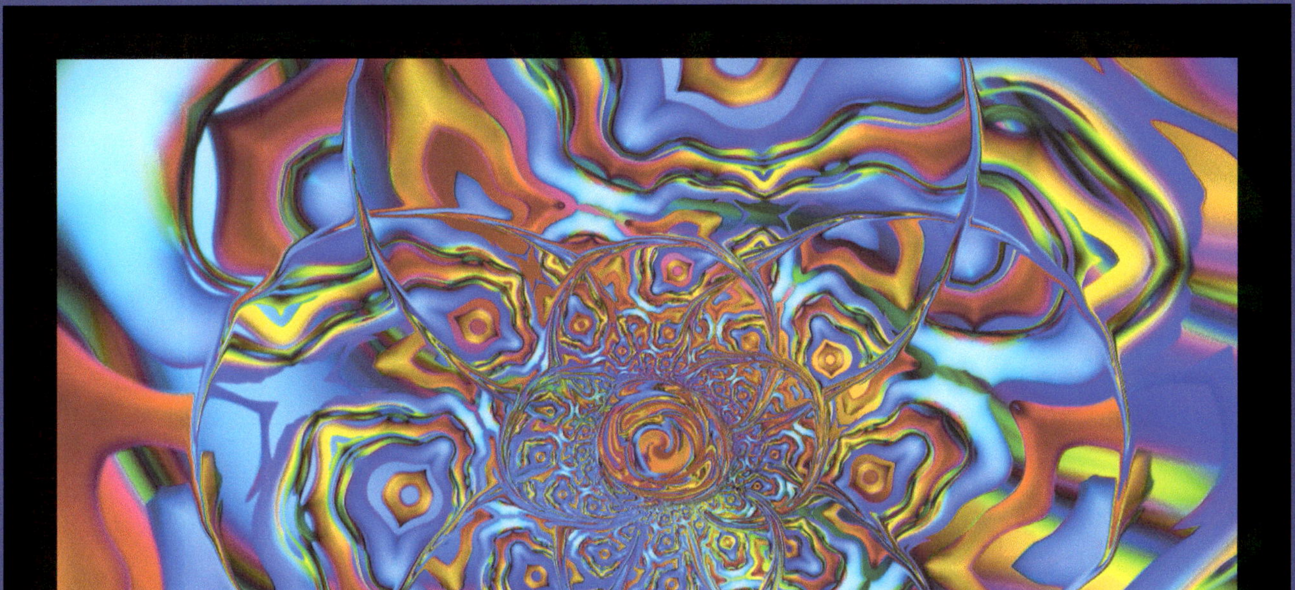

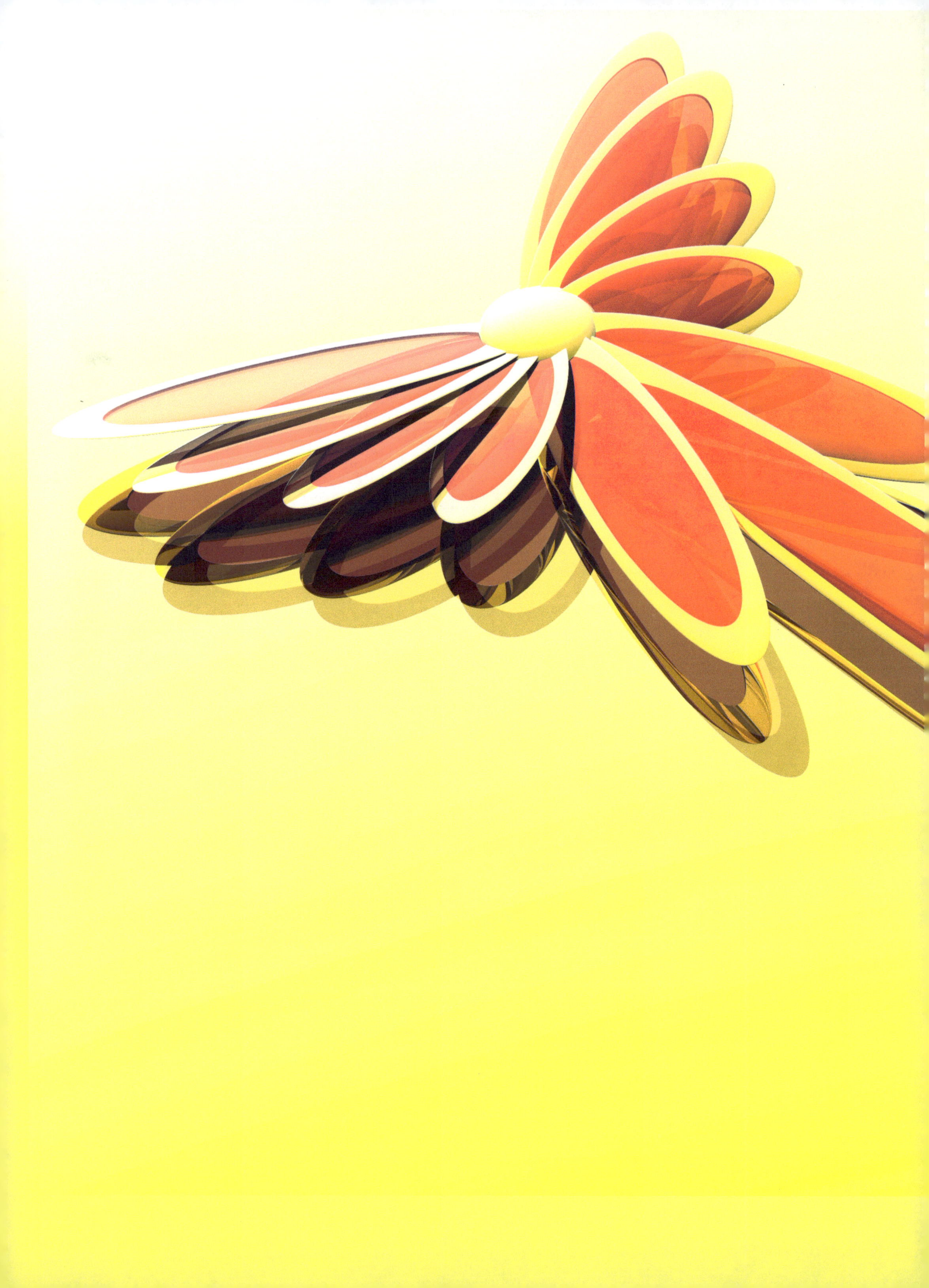

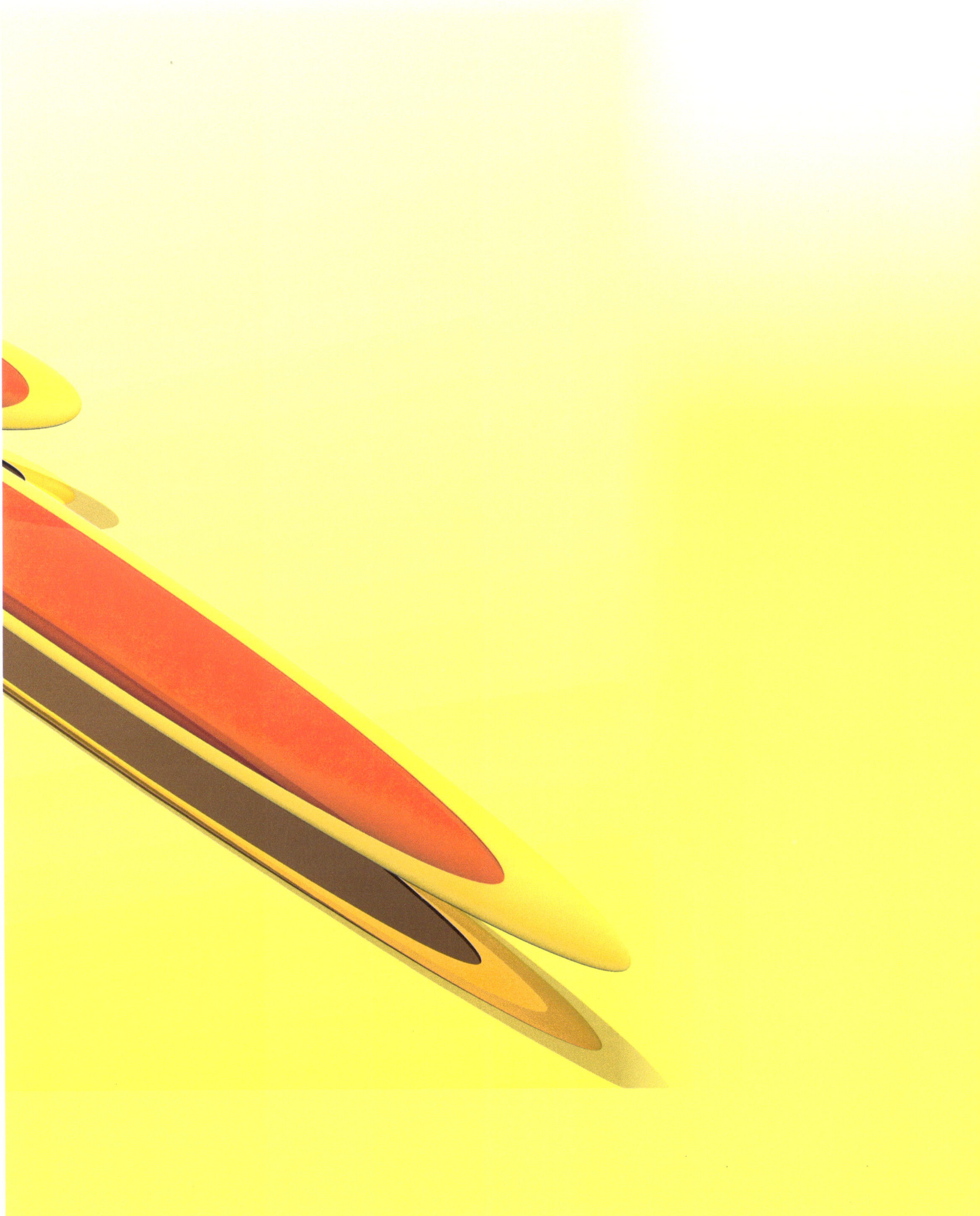

Previous pages:

GOLDWING

FRACTALSMOKE

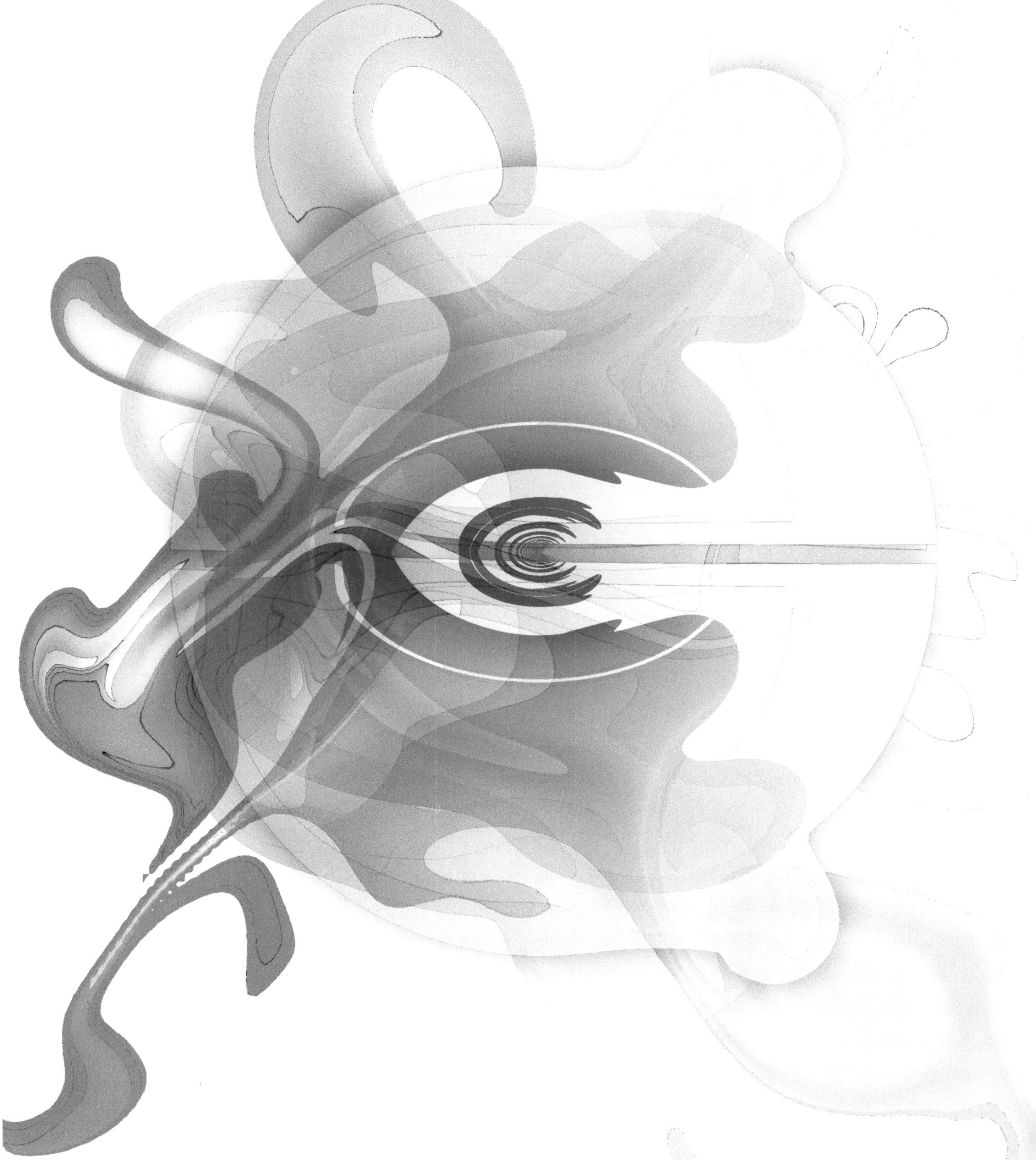

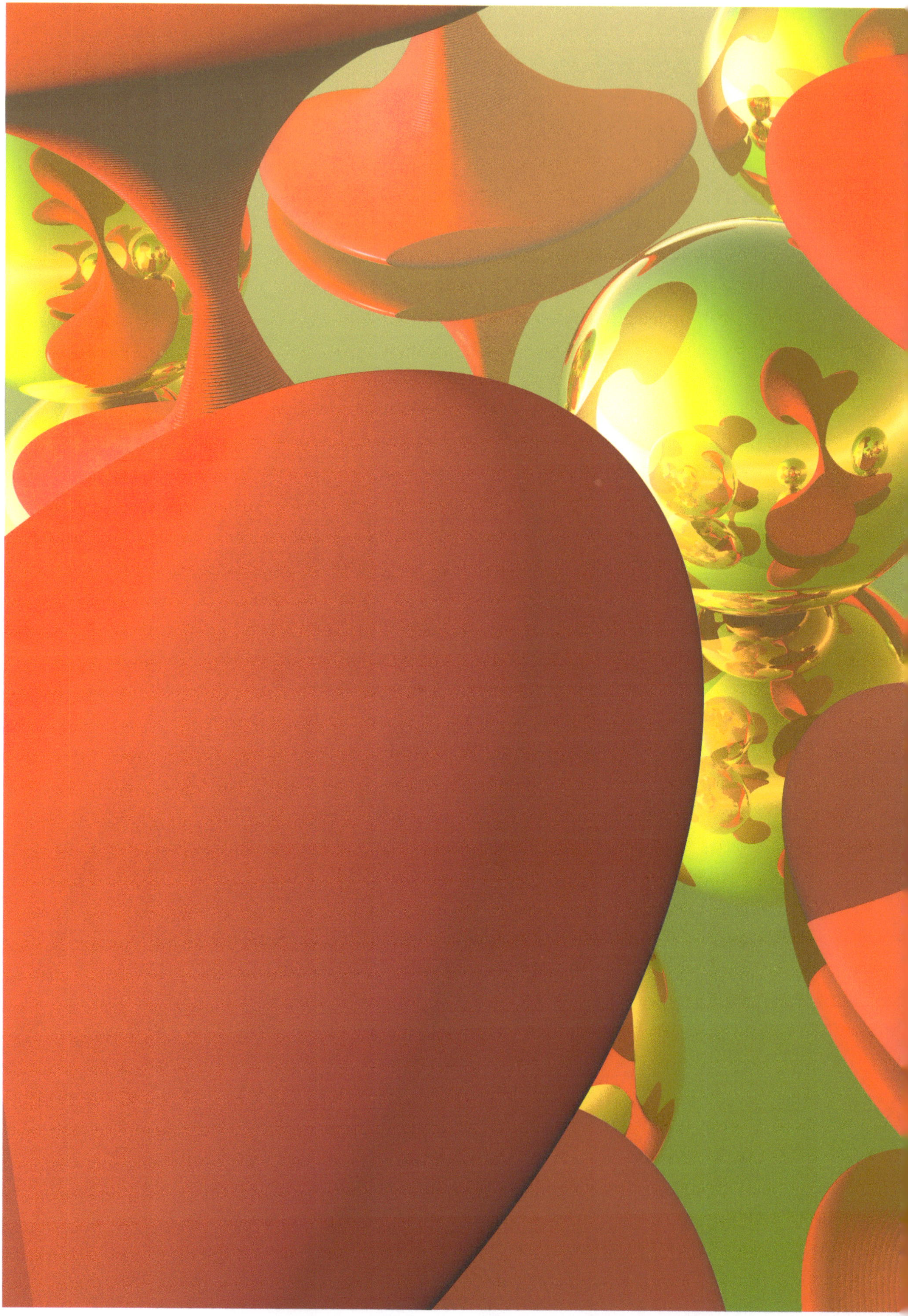

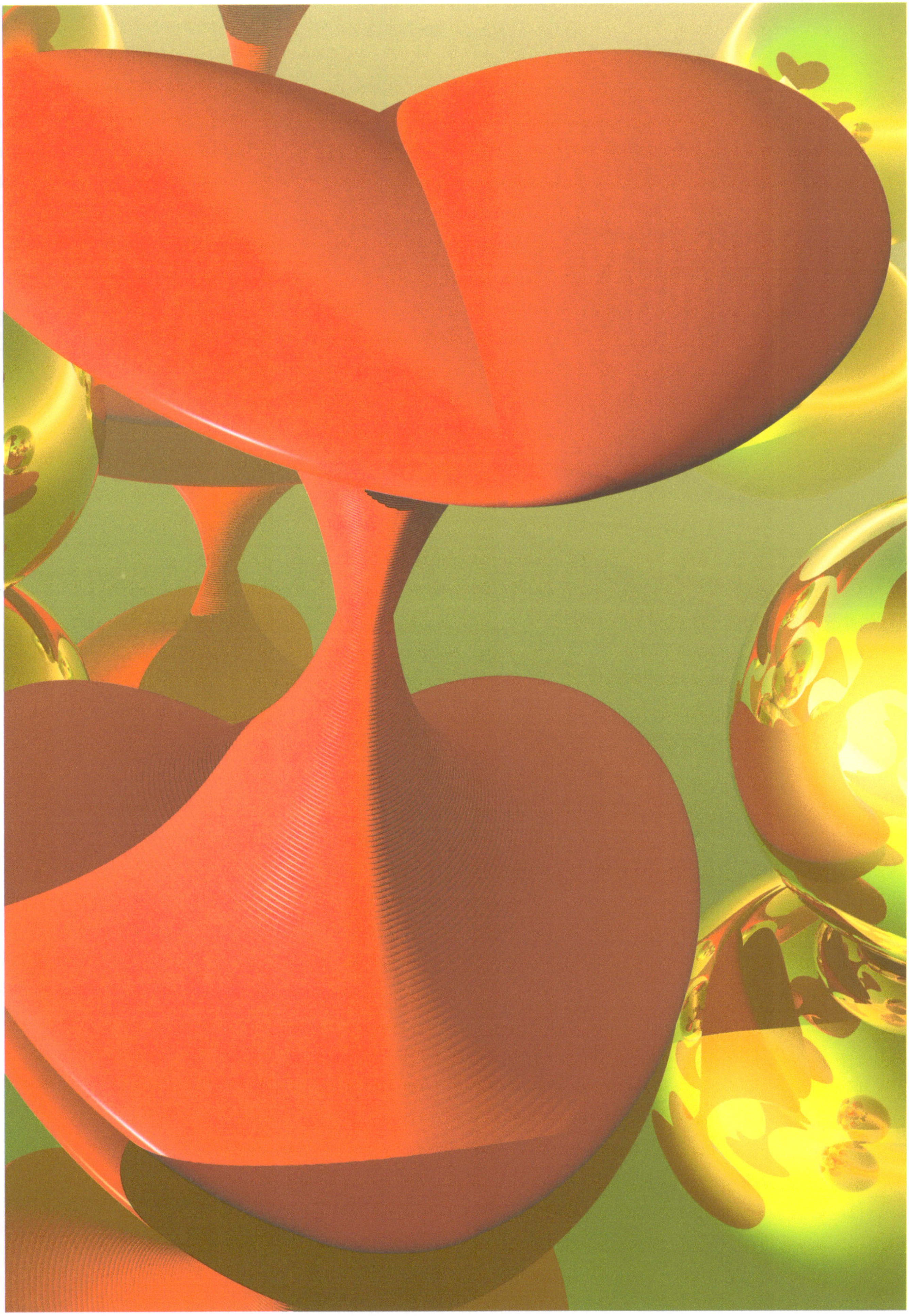

Previous pages:

DEAL OF HEARTS2

WHOAMAN

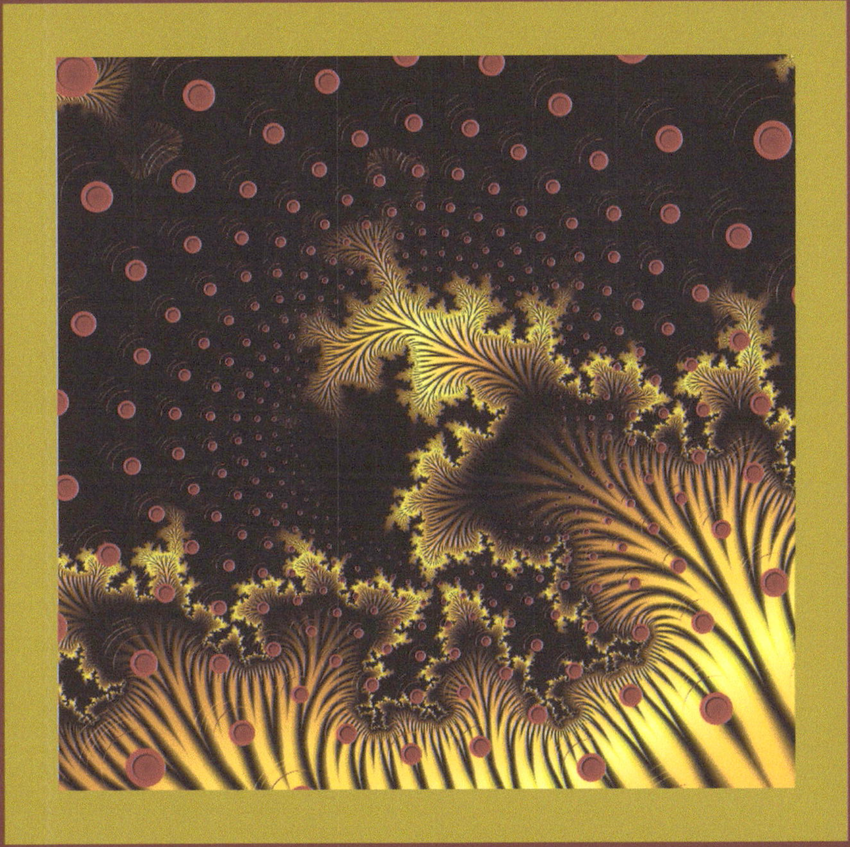

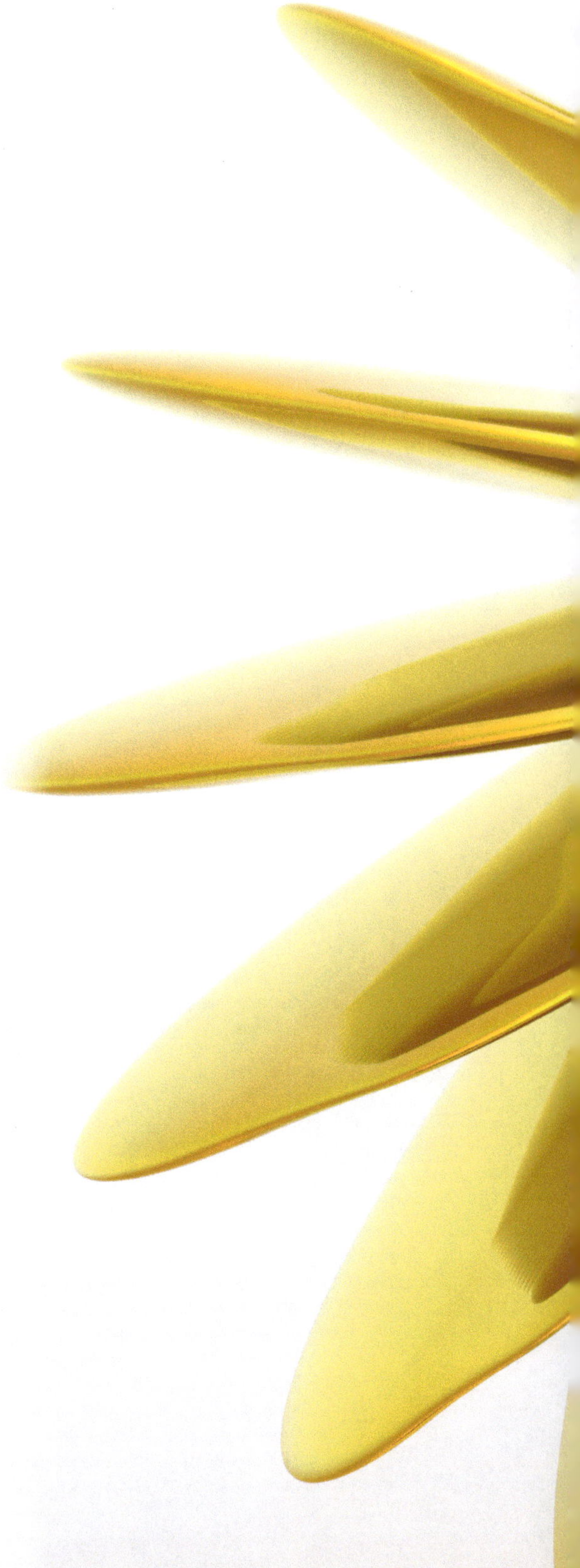

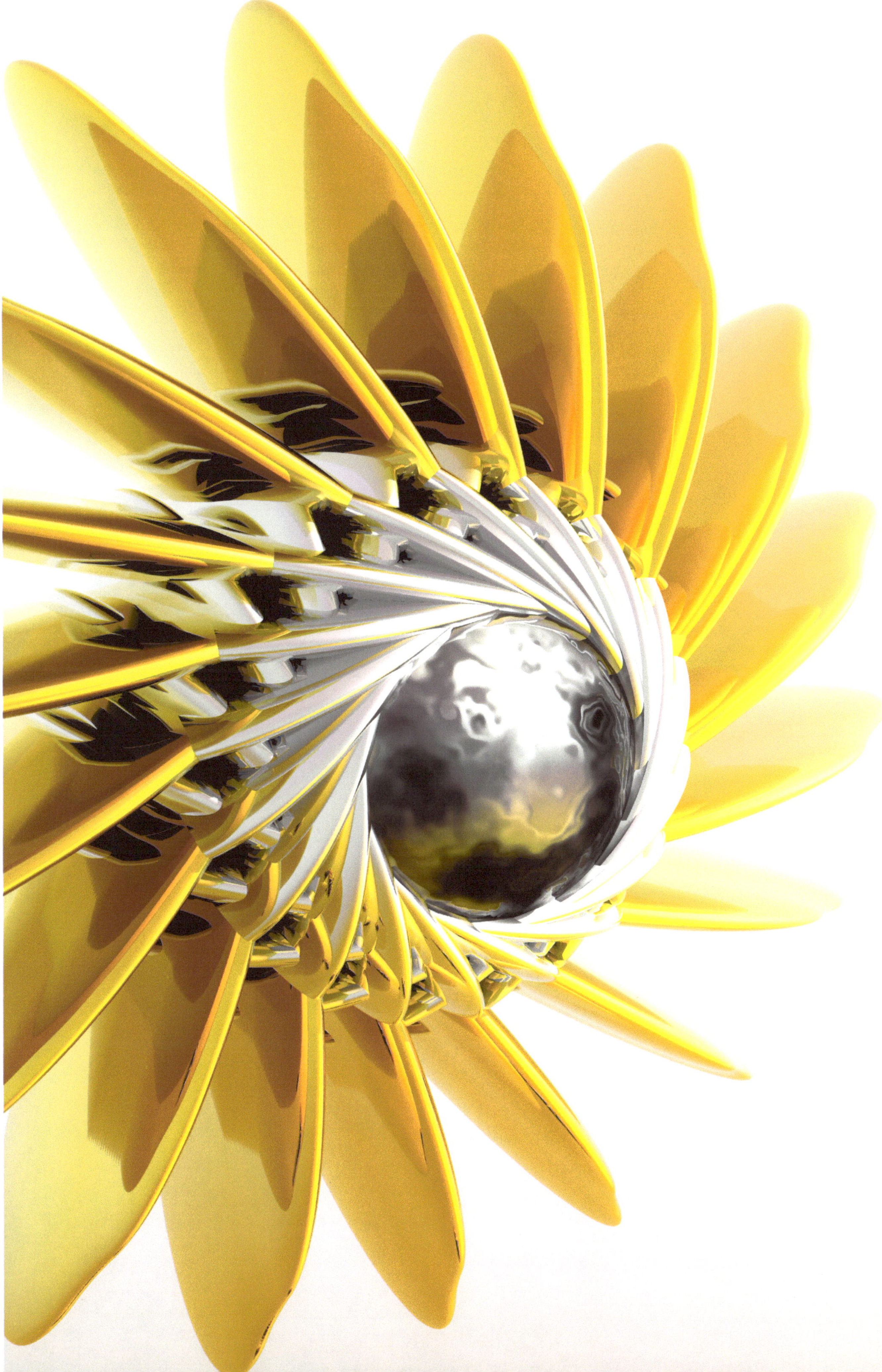

Previous pages:

VECT

PETESCATHEDRAL

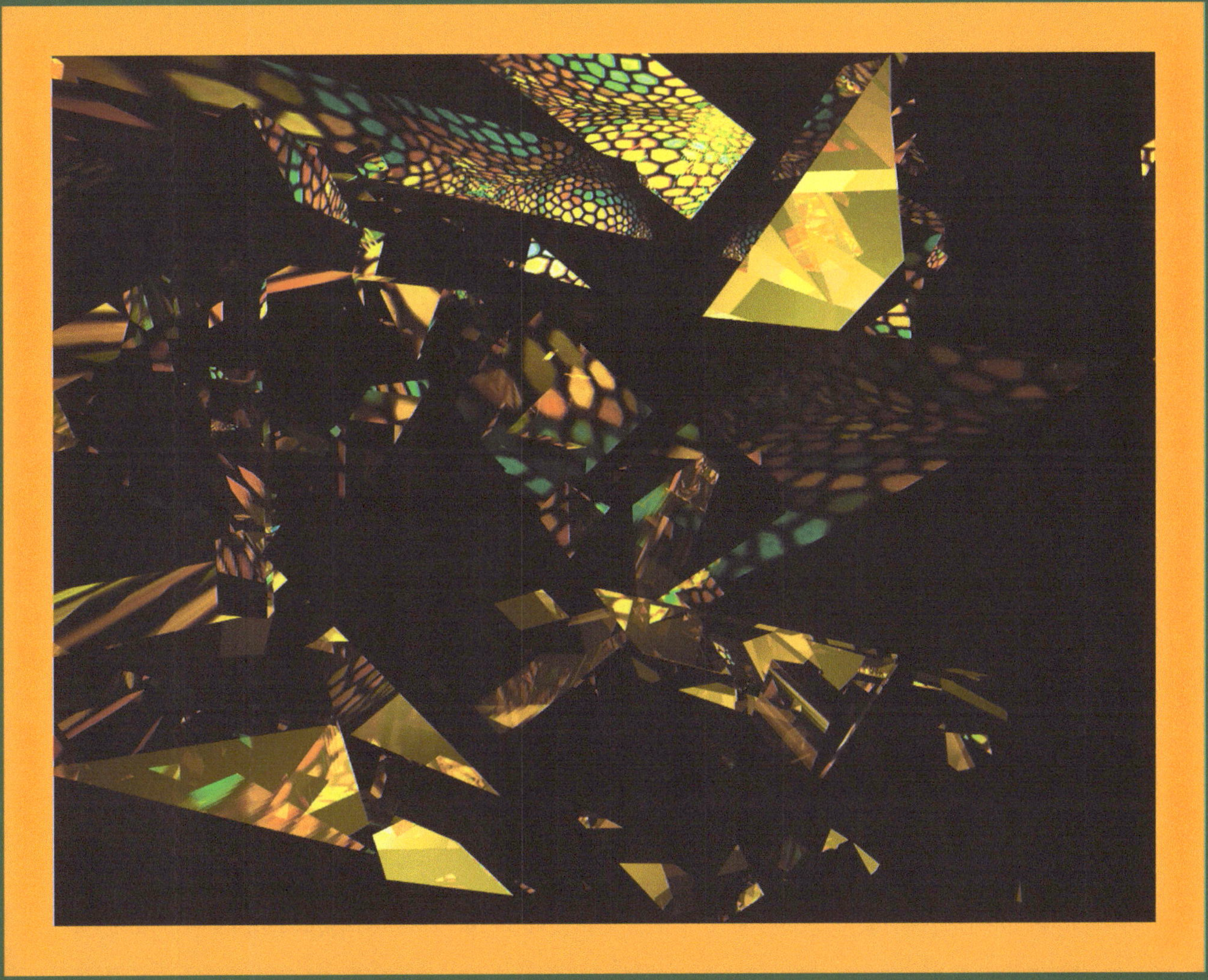

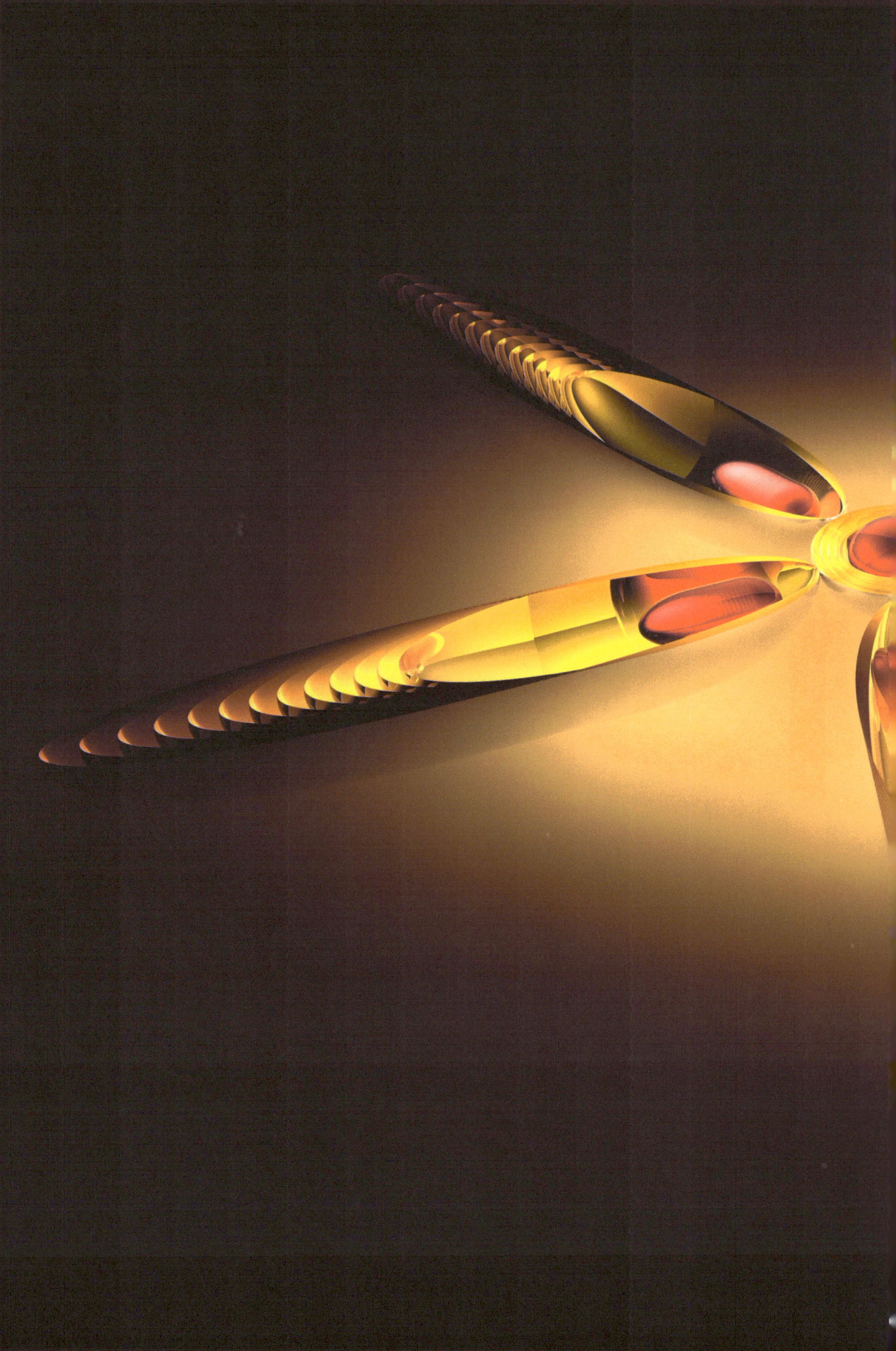

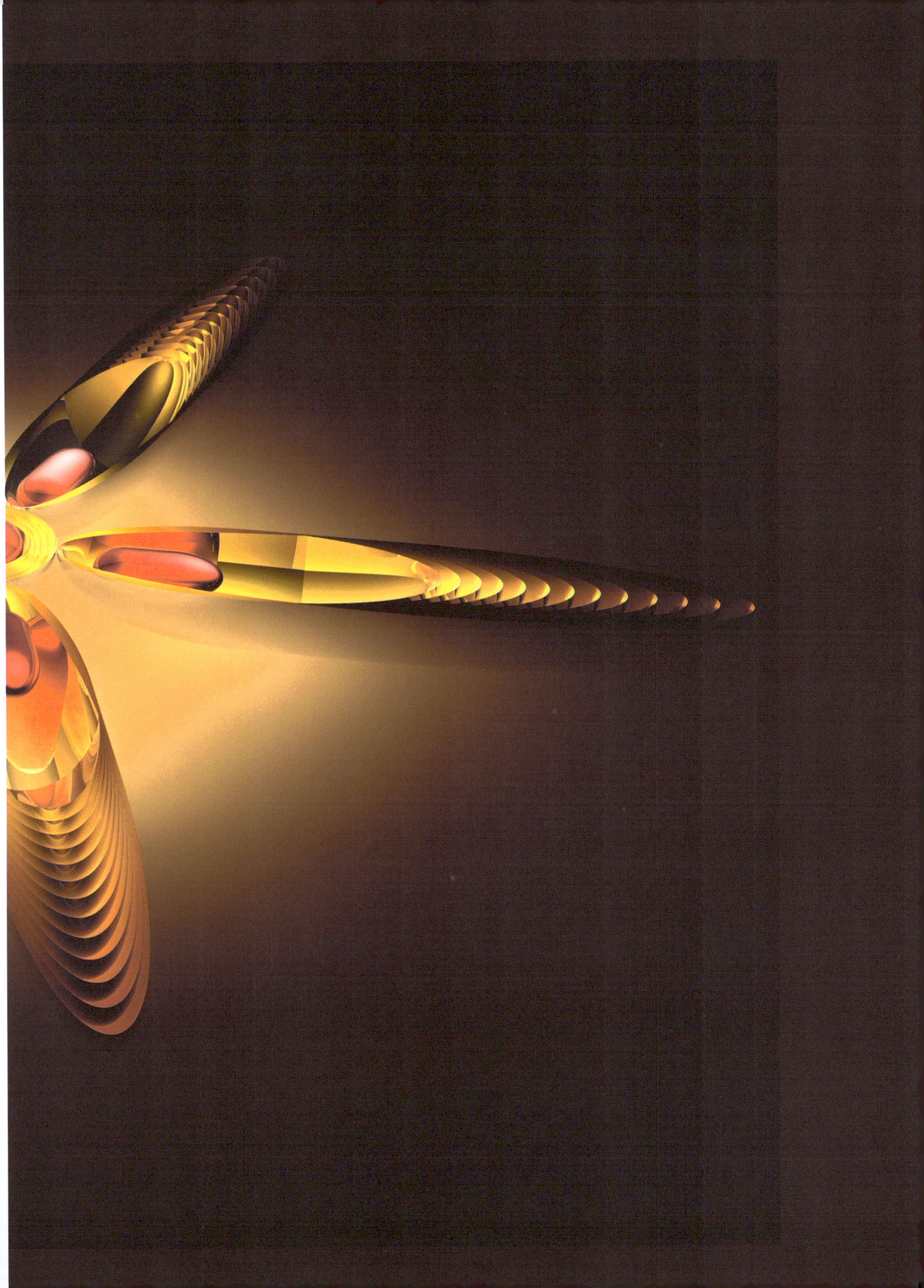

Previous pages:

DARK LOTUS

FIREBUCKET

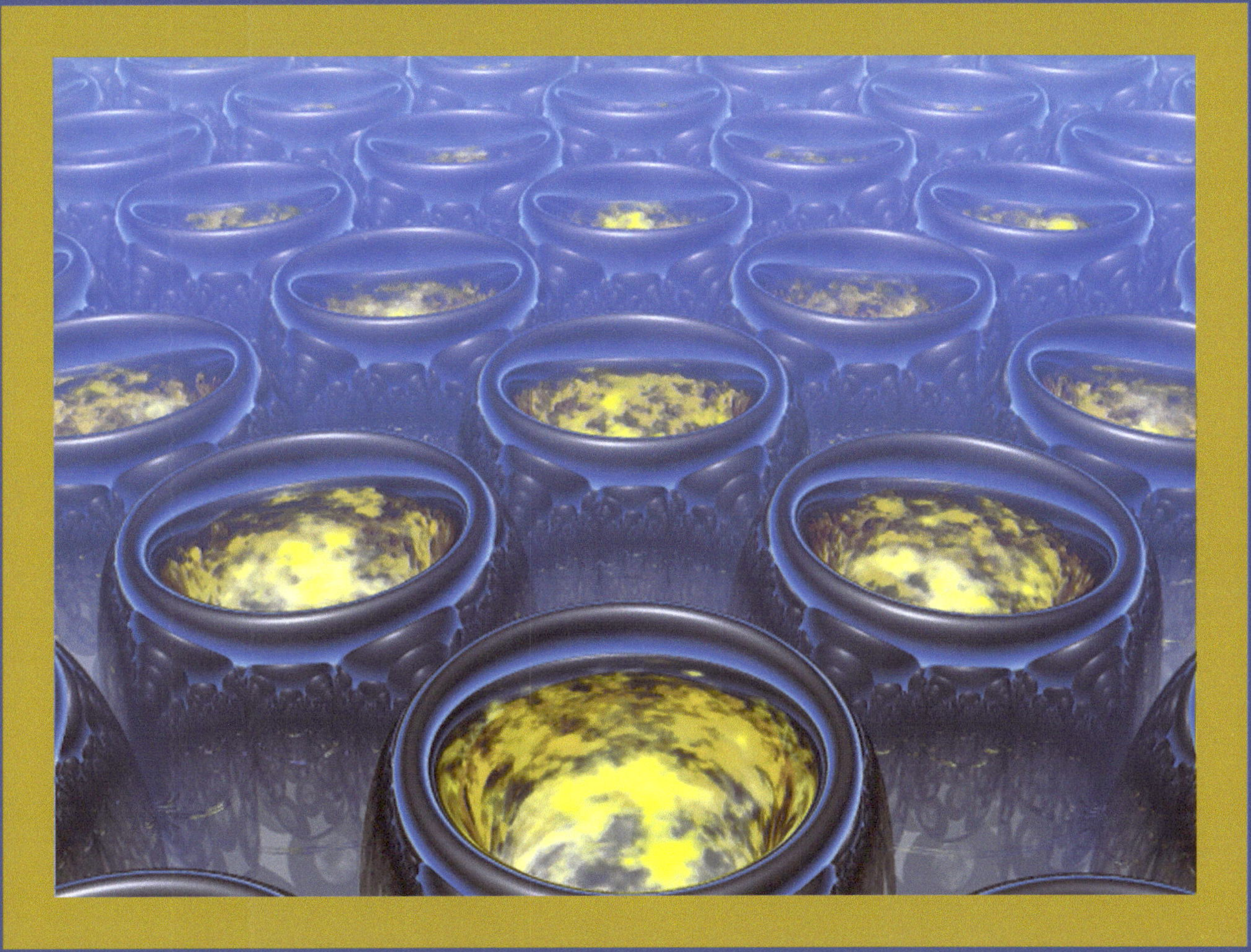

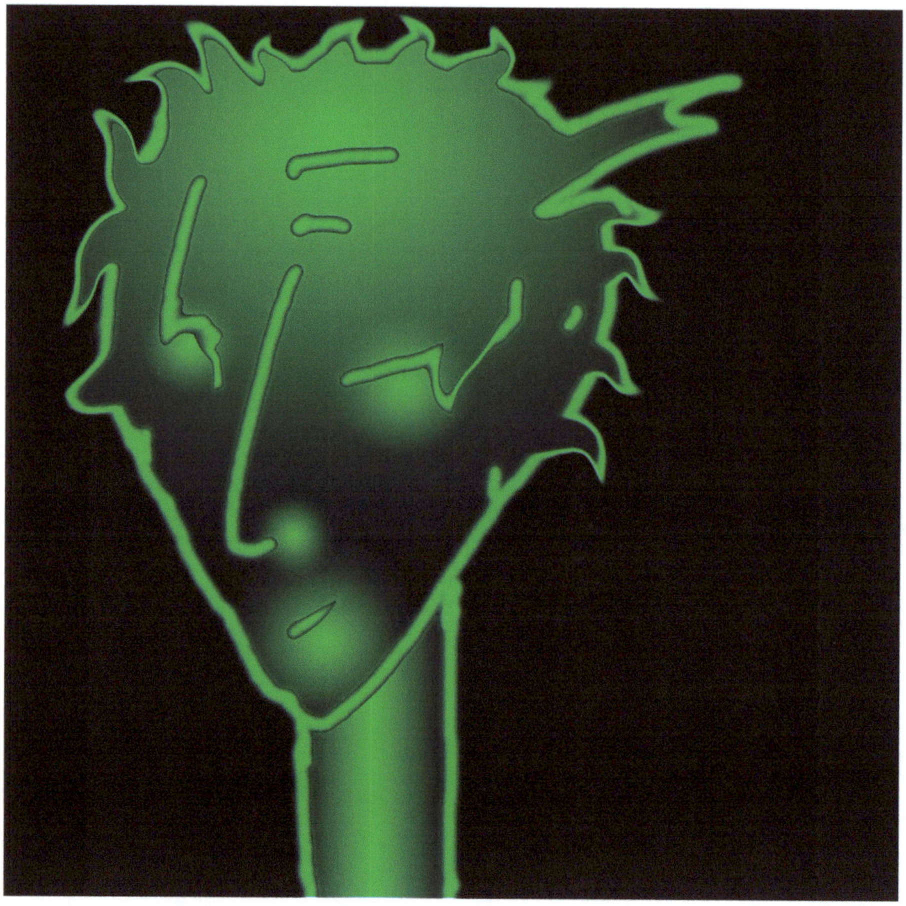

ALIENHEADZ

fractal image by Jack Cleveland

ABOUT THE ARTIST

Jack Cleveland, a polymath, had an artistic eye from an early age. Largely self-taught, his first intentional "work" was created by hammering together three pieces of driftwood to make a duck in flight; a self-portrait of himself fishing was done around age four, followed by many detailed pictures of space-ships and fantastical landscapes. He grew up on a farm in Kansas, and explored the fields and streams around his home, finding special rocks and fossils. In college, he discovered the wonders of geology and astronomy. He was greatly influenced by the images coming from the Hubble space-flight, and soon was incorporating that vision into his art-work, now done with computer graphics as he experimented with new techniques and software just emerging. An early adopter of computer technology, his fractal artwork combined his life-long love of nature and technology, brought together in this collection of original fractal art images.

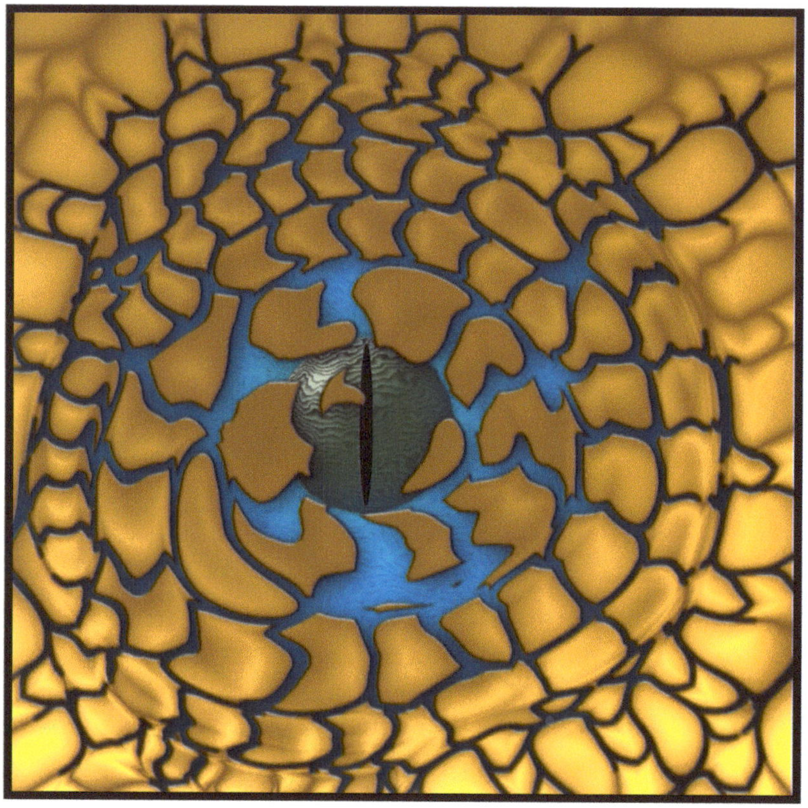

Foreword by Pat Cleveland.
Book and book jacket design by Maureen Carroll.
This book was printed and bound in the United States.

ANAMCARA PRESS LLC
P.O. Box 442072, Lawrence, KS 66044
https://anamcara-press.com